AF373954

COOLING OUR ENVIRONMENT

AN ARCHITECT'S VISION FOR COMBATING GLOBAL WARMING

GREEN BUILDINGS, SUSTAINABLE PRACTICES AND MORE

KALPANA SUTARIA
ARCHITECT

atmosphere press

© 2024 Kalpana Sutaria

Published by Atmosphere Press

Cover design by Felipe Betim

No part of this book may be reproduced without permission from the author except in brief quotations and in reviews.

Atmospherepress.com

COOLING OUR ENVIRONMENT

We can't air-condition our way out of the warming that has already happened and the warming that is coming!

Buildings are places for comfort and joy. Buildings account for more than 40 percent of all emissions and have a huge impact on our environment. Professionals in the building industry are implementing sustainable building practices to create green buildings and infrastructure to lower emissions. But for cooling our environment, we need to learn about the root causes of global warming.

All our activities continue to warm the planet. Extreme weather events are on the rise, and heat-related deaths and illnesses are going up. We witness such weather routinely. What shall we do? Personal actions are not enough, but they are motivators for bigger actions. For that, we must reach out to our lawmakers for policies that would help us transition toward clean energy research, development, and deployment. Let us focus on thoughtful policies to create a livable world for all.

PRAISE FOR
COOLING OUR ENVIRONMENT

"Cooling our Environment: An Architect's Vision to Combat Global Warming by Kalpana Sutaria is a timely and insightful exploration of how the architecture and building industries can play a critical role in addressing the global climate crisis. With over 40% of all emissions coming from buildings, Sutaria argues that the way we design, construct, and operate buildings has a profound effect on both the environment and human health. But rather than simply presenting this problem, Sutaria offers a clear vision for how we can create a more sustainable and livable world through thoughtful design and proactive policies. The book is not just a critique of current practices but a call to action."

-Charnjit Gill, author of *Pray Tell*

*For my grandchildren, your children and grandchildren,
and young people around the world*

CONTENTS

PREFACE

One of my most vivid memories of the COVID-19 pandemic was seeing footage of the majestic Himalayan Mountains, their peaks visible from great distances for the first time in many years. The sudden drop in air pollution from driving, flying, and other activities during the lockdown had led to clearer skies and breathtaking views. You may remember that moment wherever you were in the spring of 2020. Pollution affects everyone, even if you don't drive or fly. What did we learn? When we reduce the burning of polluting fossil fuels—that is, oil, gas, and coal—the air quality is improved tremendously. Unfortunately, these improvements were short-lived! Years of burning fossil fuels have collected polluting gases in the atmosphere, and that is not going away unless we take meaningful actions. Until then, the globe will continue to warm.

We see this global warming play out in extreme weather events, as well. Winter Storm Uri, which hit Texas in February of 2021, was frightening for millions of Texans who lost electricity and/or water for several days. There was no warning given by the agencies that oversaw power supply to customers. Some were deprived of medical care and suffered extreme physical and mental pain. Some people died. Many homes were damaged by plumbing breakdowns, and people had to move out of their homes. In 2017, flooding caused by Hurricane Harvey in Houston and parts of Texas displaced thousands from their homes. The impacts of such disasters around the globe are stark and affect people emotionally, physically, and financially, especially the underprivileged and vulnerable.

Professionals like me who plan, design, and construct buildings and infrastructure have relied upon oil, gas, and coal and the use of technological advancements to enhance the experiences of end users. We have recognized that the same fossil fuels that have modernized the world also cause environmental pollution. In fact, three decades ago, green builders started addressing energy efficiency, energy and water conservation, construction waste management, and other ways to lower environmental impacts. For many years, we have been implementing sustainability in building projects. That has saved resources and helped lower global warming pollutants. But global temperatures are still rising, and the year 2023 was the hottest year since 1850, exceeding the previous record set by 2016.[1] Climate scientist Zeke Hausfather of Berkeley Earth used the expression "Gobsmackingly Bananas" for the astounding hot temperatures of 2023.[1] May of 2024 extended a record-breaking temperature streak for the twelfth straight month.

The intensity of extreme events has gone up so substantially in the twenty-first century that it requires our urgent attention and implementation of thoughtful actions to address the climate crisis.

What should we do to cool our environment? Scientists continue to monitor the changing climate conditions, prepare models, and provide facts on global warming and possible adverse impacts. Here in the United States and in countries around the world, actions are being taken to transition to using clean energy sources, specifically wind and solar power. In fact, renewable energy expanded fifty percent faster in 2023 compared to 2022, according to the International Energy Agency.[2] Many scientists now believe that to avoid destruction caused by extreme events, we must build sustainably, transition to clean energy sources, and remove already collected harmful pollutants from the air.

Sustainability in buildings is a starting point. We must forward those practices wholeheartedly in policy changes needed to remove carbon from other sectors, such as manufacturing, transportation, energy generation, and agriculture.

As planners, designers, builders, manufacturers, and suppliers, we must command the attention of lawmakers, who have the responsibility to care for the well-being of their constituents by enacting policies. One important action is to elect lawmakers who are willing to listen to what the majority of Americans want.[3] We can vote in local and national elections and relentlessly continue to reach out to lawmakers until comprehensive policies to reduce global warming are put in place.

The United States Green Building Council is working in the U.S. and around the globe to promote green building rating systems that help lower carbon pollution. The American Institute of Architects and many other professional organizations have made commitments to address the climate crisis. I am a member of both these organizations. While working with the advocacy group Citizens' Climate Lobby (CCL), I understood the complexity of policymaking in the U.S. and in Texas. CCL volunteers reach out to lawmakers so that they will listen to world-renowned scientists, economists, and other experts on the facts of climate change and ask for policies to address climate change. It can be done, and it must be done.

We have not inherited this land from our ancestors.
Rather, we have borrowed it from our children.
—Kenyan proverb

AHMEDABAD, INDIA, TO AUSTIN, TEXAS

If you have experienced the hot climate of Ahmedabad, India, or any other hot place, you know the debilitating effects of heat day after day, especially if you grew up in Ahmedabad with no air-conditioning or fans to help you cope. My grandfather was a disciplined man who believed in extreme conservation. He walked to places for daily chores in that heat and remained in good health all his life. He passed away in front of my eyes on a summer day after returning from an evening walk. The heat was a contributing factor to his sudden death.

On May 29, 2024, cities in northern and central India witnessed high temperatures breaking all-time records, especially nighttime high temperatures. Cooling at night was a distant dream! With global warming on the rise, billions of people continue to suffer from the effects of heat as well as extreme weather events. The Intergovernmental Panel on Climate Change (IPCC)[4] is made up of highly qualified scientists from nations around the world to study climate change. It has published six assessment reports showing an increased intensity of extreme weather events and how they have been devastating lives, livelihoods, and shelters. It continues to raise warnings and highlight the need for actions by nations.

Ahmedabad was founded in the fifteenth century, and the fort city has homes that were built 300 to 400 years ago, long before the Industrial Revolution. These homes were built to withstand the elements as well as provide comfort using innovative building design strategies showing what is possible if buildings are designed and constructed with climate considerations.

Modern architecture developed along with industrialization that was powered by fossil fuels. New ways of design and construction were changing buildings all around the world, especially in the Western world. Daniel Barber, in his book *Modern Architecture and Climate—Design before Air Conditioning*, explores how leading architects of the twentieth century incorporated climate-mediating strategies into their designs and shows how regional approaches to climate adaptability were essential to the development of modern architecture.[5] Leading architects like Le Corbusier and Louis Khan influenced modern architecture in India, including in the city of Ahmedabad.

In the West, the use of air-conditioning started in the 1930s and 1940s and continued to increase in the 1950s and 1960s. Technological developments in lighting, cooling, and heating devices began increasing a building's comfort, increasing the environmental impacts of buildings. The green building movement of today's time emerged after the oil crisis of the 1970s. The emphasis at the time was energy efficiency and a reduction of fossil fuels in buildings. But as innovations in lighting, cooling, and heating devices proliferated, the use of fossil fuels in buildings shot up across the globe. Construction of buildings, planning of our communities, modes of transportation, and changing lifestyles have had a profound impact on warming the globe.

Sustainable building practices began emerging in the latter part of the twentieth century. The United Nations, in its 1987 report, defined sustainable development as "meeting the needs of the present without compromising the ability of future generations to meet their own needs."[6] It raised the question: How do we plan, design, and construct without degrading the environment to protect the health of people today and that of our children and grandchildren tomorrow?

I moved from Ahmedabad to Austin, Texas, in 1976. In Austin, old homes that were built prior to air-conditioning had design features like shaded entry verandas, window shades, attic ventilation, and trees and vegetation to cool the building. The City of Austin had been at the forefront of green building practices since the 1980s. The city-owned utility, Austin Energy, implemented a green building program and guidelines for energy efficiency and conservation starting in 1991.[7] As part of my work as an architect and project manager, I had an opportunity to participate in drafting the first set of sustainable guidelines for our city's bond-funded public facilities and renovation projects.

By sharing these experiences, I will show how the implementation of such policies requires persistent efforts and time at the local as well as state and federal levels to enact them. I share how professionals in the building industry and organizations not only build green projects but also need to work with environmental advocacy groups to push for changes in local, state, and national policies. These policies are key in helping the U.S. provide global leadership in the implementation of sustainable developments that would go a long way in addressing the climate crisis.

Lawmakers live among us. They are also watching adverse impacts on their constituents, the people they serve. In a democratic society like ours, we have a voice. We have many ways to demand policies. Professionally and personally, it starts with an inner will, and then your voice and your actions can start influencing decision-makers.

Young people across the U.S. and around the world are energized. They want to bring about change. I am optimistic that we can restore the environment and create a livable world if we continue to press for the right policies and work with other nations. What is better than coming together to heal and protect our people and planet? Why not take part in creating a world to enjoy with our loved ones?

In Mahatma Gandhi's words, "If we could change ourselves, the tendencies in the world would also change. As a man changes his own nature, so does the attitude of the world change toward him. We need not wait to see what others do."[8]

CHAPTER 1
EXTREME WEATHER EVENTS

OUR WORK AND EXTREME WEATHER

Construction is a step-by-step process that takes a long time and huge amounts of effort, but buildings and infrastructure can be wiped out in a short time when the worst impacts of a changing climate devastate them. We have seen images of raging wildfires and rushing floodwaters happen so quickly that people in their pathways have little time to protect themselves. Therefore, destruction from extreme weather events—hurricanes, floods, tornadoes, drought, wildfires, sea level risings—hits those of us directly involved in the building process so hard.

The energy needed for building construction comes primarily from coal, oil, and natural gas. Professionals have become acutely aware of extreme weather conditions and the polluting emissions from construction activity. They have developed a variety of sustainability rating systems and methods to measure the impacts of construction on the environment for an accounting of total carbon emissions in the entire life cycle process, from the mining of materials to the end of the building's usefulness.

While coal, oil, and natural gas have energized innovations, they have increased the use of these fuels and have also contributed to warming the planet. When we have a fever, we monitor the body's temperature and do everything to bring it down. In the same way, both our local and global concerns and commitments need to focus on warming trends with upward trajectories and take action to cool them. My goal is to reach your hearts and minds and inspire you to join me and many others who strive to secure a livable world for all.

CLEAN AIR AND WATER

The United States is the most powerful country on Earth with vast resources—natural, technological, and human—and could have a positive and far-reaching influence on improving the global climate. Fifty-three years ago, when concerns about clean air and clean water were rising, the U.S. government established the Environmental Protection Agency (EPA) to protect these basic rights for its citizens. The National Oceanic and Atmospheric Administration (NOAA) was created within the U.S. Commerce Department. NOAA supplies environmental information products, provides environmental stewardship services, and conducts applied scientific research. Using its research and analysis, NOAA provides custodianship of our environmental resources to protect the health and well-being of individuals and to help professionals and business leaders make better decisions.

In the folktale about an old man who planted a mango tree, he knew that the fruit might not bear in his lifetime but that it would benefit his children and grandchildren for generations to come. It took many years after the Industrial Revolution to understand the consequences of burning fossil fuels. Today's actions to lower harmful emissions will benefit humanity many years from now. As we gain knowledge of our environment, we use our collective skills and wisdom to address the concerns about clean air and clean water.

NOAA continues to analyze the weather and climate events and provides data on a regular basis. According to NOAA, in 2021, twenty climate disasters affected the U.S., with losses exceeding $1 billion each.[1] Overall, these events resulted in the deaths of

688 people and left significant economic, emotional, and psychological impacts. Thanks to NOAA's valuable recordkeeping, we have the frequency of disaster events from 1980 to 2022.

ANNUAL AVERAGES FOR U.S. CLIMATE DISASTER EVENTS

- 1980-1989: 3.7 events on average per year
- 2013-2022: 18.1 events on average per year

While this data only accounts for the U.S., similar weather events, including rising sea levels, extreme heat, untimely and heavy rains, flooding, droughts, wildfires, crop failures, power loss, cyclones, melting glaciers, and winter storms, are rising in every part of the world. In Africa's countries, for example, agencies like NOAA do not exist. In many countries, weather and climate records are not easily available, but conditions on the ground speak for themselves. It makes no sense to continue this path or allow for unnecessary losses of lives and billions of wasted dollars.

INCREASING HEAT

NOAA keeps records of temperatures and effects of warming, using all the available resources since the Industrial Revolution, and provides analysis of this data for use. In Berkeley Earth's analysis, the global mean temperature in 2023 is estimated to have been 1.54 ± 0.06 °C (2.77 ± 0.11 °F) above the pre-industrial average temperatures. This is the first time that an annual average temperature has exceeded the pre-industrial baseline period by more than 1.5 °C (2.7 °F).[2] According to NOAA, this extra heat is driving regional and seasonal temperature extremes, reducing snow cover and sea ice, intensifying heavy rainfall, and changing habitat ranges for plants and animals—expanding some and shrinking others. Land, oceans, and the Arctic regions are warming, but the Arctic region is warming at a much faster rate.[3]

Figure 1 shows a stark representation of the warming trajectory, especially from 1970 onward. Variations occur from day to day, month to month, and year to year, but what matters is the average overall increase in heat and the effects of accumulated heat on our environment.

Notice the sharp uptick on the graph showing the hottest temperatures ever recorded in 2023. *The Hill* reported that 2,325 people were killed by heat in 2023. According to Berkeley Earth, the average temperature rise crossed 1.5 degrees Celsius, which is set as a danger threshold by the Intergovernmental Panel on Climate Change (IPCC).[4]

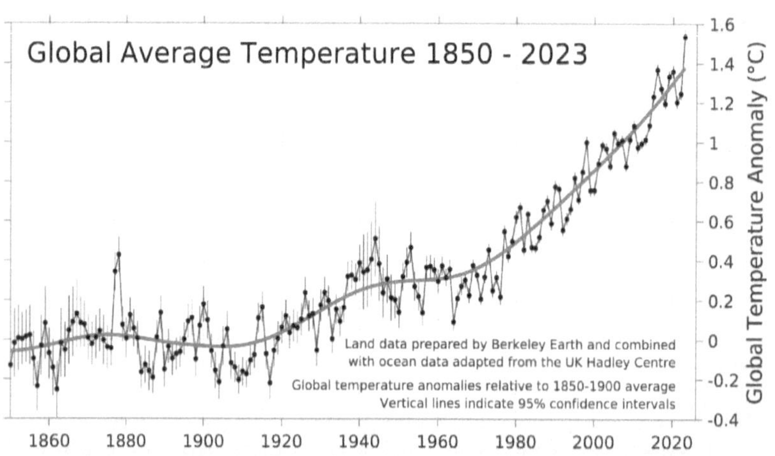

Fig. 1. Global temperatures rise in 2023.
Berkeley Earth (berkeleyearth.org/global-temperature-report-for-2023/)

HEAT AND DROUGHTS

In my adopted hometown of Austin, Texas, we have had similar warming rates as the rest of the globe. Drought conditions prevailed along with record-breaking temperatures in 2022 and 2023.

Climate Central, a non-profit climate-science analysis and reporting organization, found that central Austin is more than seven degrees Fahrenheit warmer than its surroundings due to the urban heat island effect.[5] As professional planners, designers, and builders, we need to study the comfort of users while planning for new buildings to ensure that all possible design features are used to cut down on rising utility bills for cooling and heating. For existing buildings, we need to add energy efficiency features to reduce cooling or heating needs, with improvements in local regulations and building codes.

The effects of heat are starkly visible in Seattle, Washington, home of my grandchildren. Weather in Seattle used to be pleasant most of the year, and the city's use of air-conditioning equipment (A/C units) was not standard practice, but residents have started adding them.

My hometown of Ahmedabad, India, was always hot. It is now on a rising curve like the rest of the world. A/C units were rarely a part of building construction due to affordability. Buildings were constructed to provide needed ventilation for user comfort to the extent possible. Affluent people have started installing A/C units in their bedrooms and living rooms. Ahmedabad was so hot in the summer of 2022[6] that one of my relatives described it this way: "Weather here is 110 degrees [Fahrenheit] most days. My A/C is on 24/7, so we are doing fine. We still go out for dinner in an air-conditioned car to air-conditioned restaurants."

However, many in Austin, Seattle, and Ahmedabad are not able to afford, install, or operate A/C units on a 24/7 basis. For them, adaptation is the only option. Vulnerable people suffer from heat-related health issues. On October 20, 2022, the air quality in the city of Seattle was worse than in Beijing, China, or Delhi, India, where air quality is routinely harmful. Why? Forest fires raged in the Cascade Range mountains during weeks of unusually dry and hot weather.

Smoke from wildfires makes it difficult to be outdoors. The latest U.S. wildfire smoke spread across the North, Northeast, and Midwest starting on June 7, 2023. The fires created so much pollution that authorities asked people to stay indoors to avoid breathing polluted air. Orange and hazy photos of New York City were reminders of what smoggy skies felt like in the Los Angeles of the 1970s.

The wildfires lingered for days over a large region, from Maryland to the Canadian capital, Ottawa. Wildfire smoke hit the New York area again on June 28, 2023. Dr. Stephen J. Pyne,[7] a historian with a special focus on the history of fires, told CNBC that wildfires in Canada have been serious before, but climate change is "a performance enhancer," meaning climate change intensified the wildfires. Despite all precautions taken by the people of this region, polluted air manages to affect the health of vulnerable people.

Why show the map of 2023, when we experienced twenty separate billion-dollar weather and climate disasters in 2021? In 2023, we passed that number in the eight months from January to August, with a total number reaching twenty-eight by the end of 2023. Notice the type of disasters hitting different regions vary each year. The Canadian wildfires, which polluted New York City and the Northeast in June, are not shown on NOAA's map because they originated in Canada.

HURRICANES AND FLOODS

In 2021, regions east of the Rockies had hurricanes, flooding, tornadoes, and hailstorms. The West experienced droughts and wildfires, which destroyed towns and forests, creating air quality issues that left people with respiratory illnesses. Residents in California, Oregon, and Washington suffered from unusually high heat. According to NOAA, in the fifteen years between

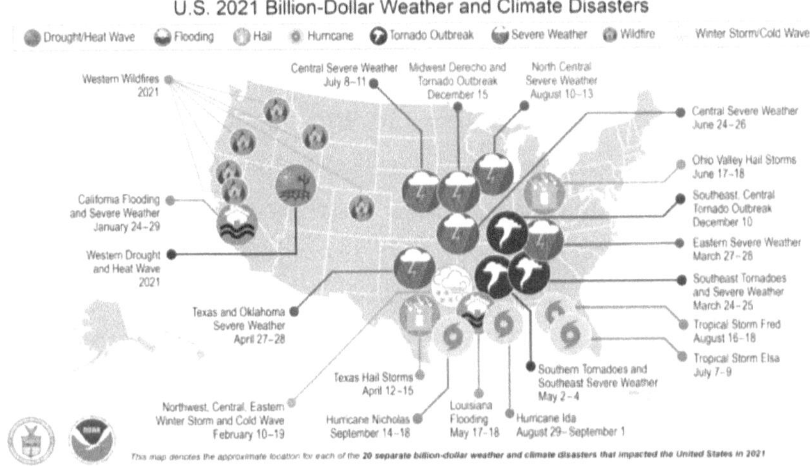

Fig. 2. National Oceanic and Atmospheric Administration (noaa.gov) shows billion-dollar weather and climate disasters in 2021. Such events have crossed this number in eight months in 2023. (Image credit: NCEI.NOAA.gov)

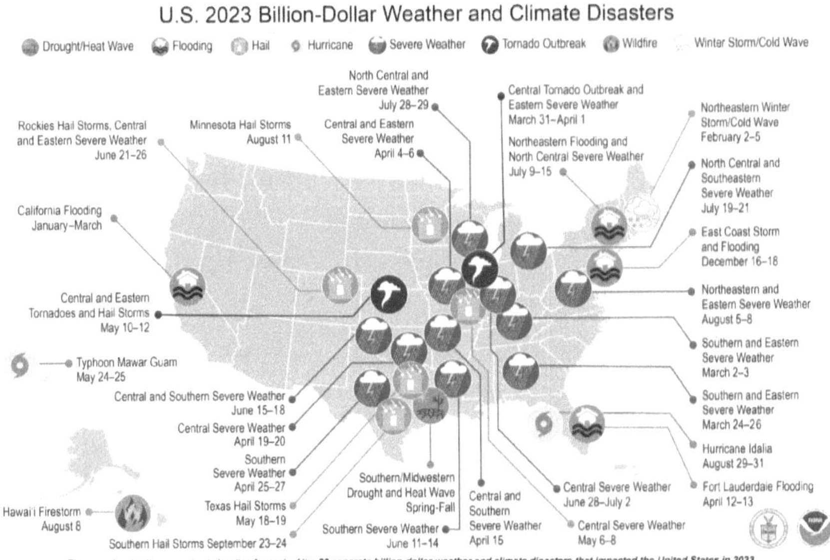

Fig. 3. A map of the U.S. plotted with 28 weather and climate disasters, each costing $1 billion or more that occurred in 2023. (Image credit: NCEI.NOAA.gov)

2005 and 2019, 156 separate billion-dollar weather or climate disasters in the U.S. have cost a combined total of $1.16 trillion in damages.[8]

Hurricane Harvey, a devastating Category 4 hurricane, made landfall in Texas and Louisiana in August 2017, causing catastrophic flooding, many deaths, and the displacement of thousands of people. The damage from Hurricanes Harvey, Irma, and Maria alone in 2017 is responsible for a loss of approximately $265.0 billion, according to NOAA.[9] The National Weather Service reported that an entire year's worth of rain, more than fifty inches, fell in one week over Houston and other regions of Texas. Following Hurricane Harvey, approximately twelve percent of buildings in Houston were flooded.[10]

Hurricane Ida, on August 29, 2021, became another deadly and destructive Category 4 Atlantic hurricane—the second-most damaging hurricane on record to make landfall in the state of Louisiana, behind Hurricane Katrina, which caused massive death and destruction in 2005. Hurricane Ian was another deadly storm to strike the U.S., mainly the state of Florida, arriving on September 28, 2022, causing many deaths and billions in damages.[11]

Even when federal aid is available, victims face the complexities of bureaucracy, insurance claims, and disruptions in living conditions. They wait for years to receive funding to rebuild their lives and shelters. Jake Bittle, in his book *The Great Displacement*, provides vivid stories of displaced people and the agony of rebuilding lives after disasters.

2022 FLOODING IN PAKISTAN

In the 2022 monsoon season, rainfall in Pakistan was nearly three times the average of the past thirty years, according to Pakistan's National Disaster Management Authority. In Sindh

Province, which borders the Arabian Sea to the south, the rainfall was nearly five times the country's average. Hundreds of people were killed or became homeless. Buildings were washed away or submerged. Roads and bridges were destroyed. Crops were wiped out. One-third of the country was underwater.

According to Dr. Steven Clemens, professor of Earth, Environmental, and Planetary Sciences at Brown University, who focuses primarily on reconstructing changes of monsoonal rainfall in the Indian and East Asian systems, the excessive rain patterns in Pakistan are "super-consistent with what we expect in the future" as the planet heats up.[12] The damages were more than $12 billion, and the rains destroyed homes, bridges, roads, vegetation, agricultural fields, lives, and livelihoods. At one point, it rained seventy-two hours nonstop. This event has been dubbed a "monsoon on steroids" or "monsoon monster."[13] People who suffered were contributing to global warming much less than those of us with cushy lifestyles in the industrial countries.

Ahmedabad, which is roughly 250 miles southeast of Pakistan, had similar rain-bomb events after the monsoon season was supposedly over. The house of an eighty-year-old relative, who has lived all her life in this city, flooded one night without any significant warning while she was sleeping. In September of 2023, cities in my home state of Gujarat, including Ahmedabad, experienced flood conditions not witnessed before. The flooding devastated millions of people in its pathway and damaged crops.

On April 12, 2023, the airport in Fort Lauderdale, Florida, received 25.91 inches of rainfall.[14] This total was roughly equivalent to a third of the city's annual rainfall and seven times the typical April total, according to the National Weather Service. The damage to lives, livelihoods, and properties is undetermined at the time of this writing.

The Federal Emergency Management Agency (FEMA) oversees the determination of flood plains and makes major revisions when necessary. It is hard to imagine how any flood control design that professionals in the building industry, who design and construct per code, could handle this kind of excessive rainfall in short periods of time. After building and rebuilding flooded properties, residents in some parts of the U.S. gave up. FEMA bought their properties, and residents moved to different locations.

THE SOUTHWEST MEGADROUGHT AND WILDFIRES

When we hear about droughts around the world and how they can cause major migrations, we must think about how global warming is going to make bad situations much worse due to water shortages and food supply. Currently, in the states of California, Nevada, Colorado, and Arizona, a severe drought is already affecting agriculture and the water supply. Other states like Oregon, North Dakota, Utah, Idaho, and Washington are also experiencing drought conditions.

The Southwest's Lake Mead is the largest reservoir in the country, formed by the Hoover Dam on the Colorado River. The reservoir's water level had significantly dropped by May 2022. The Colorado River supplies drinking water to forty million Americans in seven states as well as a part of Mexico. The river irrigates 5.5 million acres of farmland. In May of 2023, the Bureau of Reclamation, an agency within the U.S. Department of the Interior, worked with Arizona, Nevada, and California to come up with an agreement to protect big cities and the farmland along the Colorado River. Water shortages throughout this region have compelled cities and states to make difficult decisions, such as the one made by Phoenix to reduce new development permits and better manage long-term water supplies.

According to NOAA, "rising temperatures, declining snowpack, and frequent droughts are all leading to a dramatic surge in wildfire frequency and severity across the western United

States. Climate change is loading the dice, transforming what was once a natural, cyclical, and seasonal visitor to the landscape into an omnipresent threat. Year after year, fire is filling Western skies with smoke, converting live timber and understory vegetation to burn scars, impairing water supplies, disrupting economies, threatening lives and property, and altering the landscape for generations."[15]

On October 20, 2022, the people of Seattle, Washington, and Portland, Oregon, suffered the worst air quality in the whole world because of wildfires. My own family members described it as the worst air-quality issue they had suffered after living there for more than twenty years. Even when staying indoors, people suffered from respiratory issues.

AIR POLLUTION AND ALLERGIES

Austin is often called the Allergy Capital of the World. It has been my adopted hometown for many years, and I have suffered from Austin's pollen allergies for a long time despite taking an allergy shot treatment, medicines, and necessary precautions. I frequently meet people with similar health issues while going through a second set of allergy shots. In fact, asthma is common among the hardest-hit patients with allergies, especially children.

Carbon dioxide (CO_2) is a colorless and odorless gas that we can't smell or see, but its levels have gone up, and it continues to accumulate in the atmosphere. The Keeling Curve represents the concentration of CO_2 in Earth's atmosphere since 1958. On June 5, 2023, the concentration rose to 424 parts per million (ppm) compared to 317 (ppm) in 1958.[16] With an increase in CO_2 levels and ground-level ozone, the combined effects of pollution and pollen allergies have become worse.

Ozone (O_3)[17] close to the ground is harmful to human health

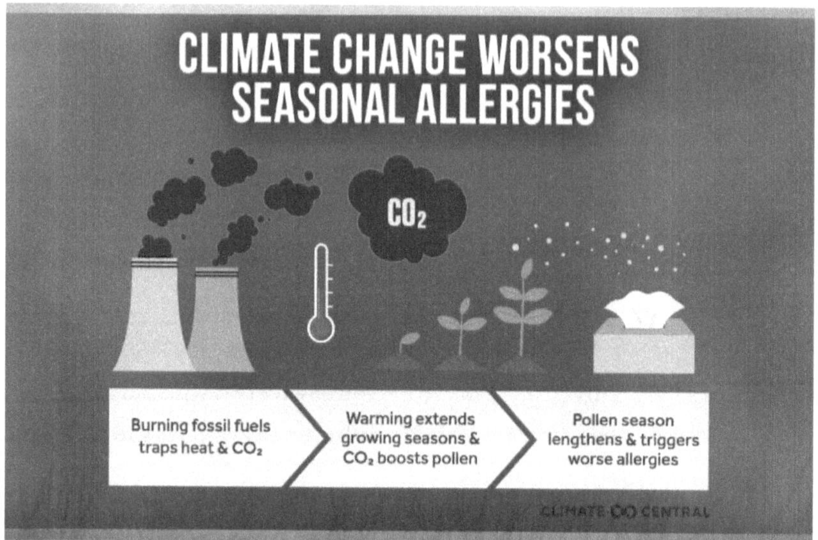

Fig. 4. Increased CO2 levels mean more pollen, affecting the health of people suffering from allergies.
Source: https://www.climatecentral.org/graphic/warming-climate-more-pollen-worse-allergies

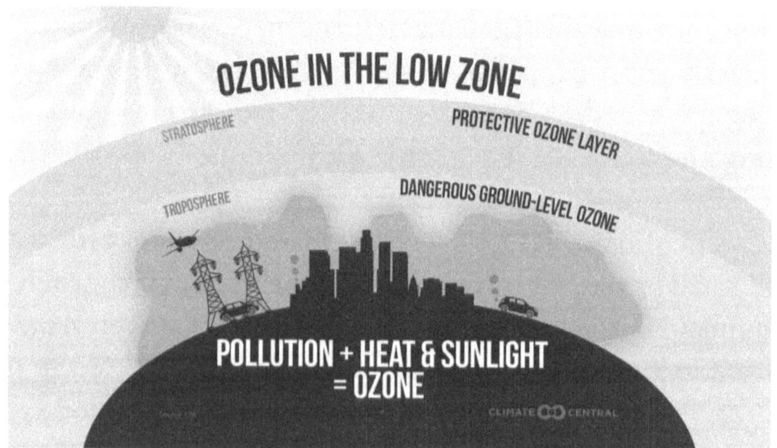

Fig. 5. Increased level of ozone due to pollution and warming.
Source: Climate Central[19]

and crops like wheat, corn, and soybeans. People suffering from pollen allergies, as I do, have an instant physiological effect when ground-level ozone increases along with pollen from ragweed, grass, mold, trees, and other pollutants. Sneezing, headaches, postnasal drainage, and a general sense of tiredness are the symptoms that I suffer from even when staying indoors. Austin, in 2022, had twenty-five Ozone Action Days as of October 15, which topped the combined total of the previous eight years.[18]

In India, global warming and increased tailpipe pollution have made air quality so bad that all of the country's major cities frequently have levels deemed unhealthy. In 2019, I visited New Delhi for a short time, but breathing polluted air affected my respiratory system so badly that I had a lingering cough for three months. Air pollution affects productivity and degrades quality of life. According to the World Health Organization (WHO), air pollution is responsible for an estimated seven million premature deaths worldwide every year.[20]

MELTING GLACIERS

The beautiful sight of the Himalayan range is unforgettable for those who have enjoyed its snow-covered peaks. I was lucky enough to visit Kashmir in the 1960s to witness them. The spread of Himalayan glaciers covers a vast area, from Afghanistan to the west and Sikkim to the east. Many mountaineers have enjoyed climbing the highest mountains, including Mount Everest. These glaciers supply drinking water to over a billion people in the region. Black carbon pollution has warming effects that are several times worse than carbon dioxide. This has accelerated the melting of glaciers around the world, including the Himalayan glaciers. The International Center for Integrated Mountain Development in Kathmandu found that glaciers in regions such as the Hindu Kush and the Himalayan Mountain ranges have melted sixty-five percent faster from 2010 to 2019 than in the previous decade.[21]

Figure 6 below shows two views of the Rongbuk Glacier, one in 1968 and the other in 2007. Rongbuk is the largest glacier of Mount Everest's northern slopes, which feeds the Rongbuk River, a source of water for millions of people. The disappearance of these glaciers will likely cause serious water shortages. Groundwater levels have been falling. My relatives in the Delhi and Punjab areas must depend upon deep wells for water supply because of the enormous growth and developments in the urban areas that are straining water supplies.

Fig. 6 - **Top.** The Rongbuk Glacier of the Himalayan Mountain range in 1968.
Fig. 6 - **Bottom.** The Rongbuk Glacier in 2007 shows the disappearance of snow.
Source: Mountaineer David Finlay Breashears[22]

Like many glaciers around the world, Greenland's glaciers continue to retreat and lose ice. According to NASA, Greenland has lost more than 5,000 gigatons of ice to the ocean. It is hard to visualize gigatons of ice.[23] To understand it, I had prepared the map in Figure 7 - Left, showing retreating glaciers from 1992 to 2007. Climate scientist Dr. Jim Hansen, in an *Ecologist* article on March 24, 2016, "Ice melt, sea level rise and superstorms:

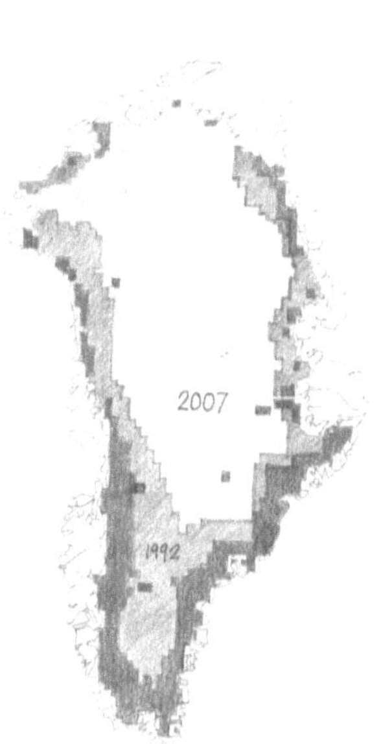
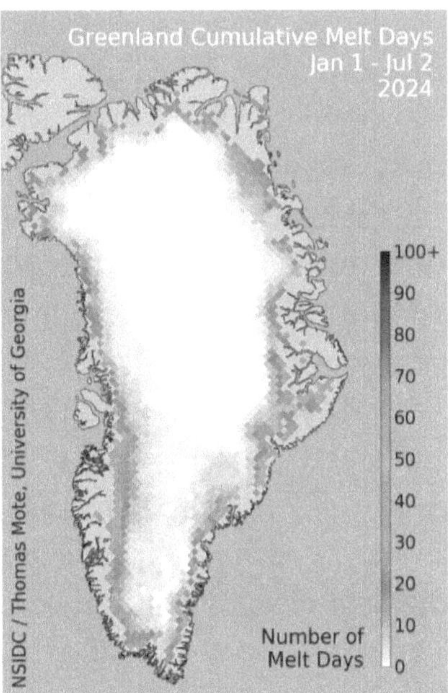

Fig. 7 - Left. Author's sketch of Greenland shows retreating glaciers in 1992 and 2007.
Fig. 7 – Right. Source Credit: National Snow and Ice Data Center, https://nsidc.org/ice-sheets-today

the threat of irreparable harm," had this warning: "If Greenland freshwater shuts down deepwater formation and cools the North Atlantic several degrees, the increased horizontal temperature gradient will drive superstorms stronger than any in modern times. All hell would break loose in the North Atlantic." Sea level rise remains a major concern as described in this article.

WINTER STORMS

Texas's Winter Storm Uri in February 2021 remains the scariest event I have personally experienced. In our Austin house, we lost electricity in the morning hours of February 15 without any notice. We didn't get the power back for seventy-seven hours and were among the sixty-nine percent of Texans without electricity in frigid conditions. Once our electronic devices used up power, we were cut off from family and friends. The snow was six to eight inches thick, and in a city without snowplows or the equipment to handle winter weather, we had no ability to go anywhere.

Thanks to having an energy-efficient home, the temperature inside remained forty-two degrees Fahrenheit. We used up most of our candles. Only later did we find out the horrors of this event. According to the KVUE 24 News of Austin,[24] the Texas Department of State Health Services (DSHS) released its sixth and final report on Winter Storm Uri, which reported deaths of 246 people because of Uri and attributed most fatalities to hypothermia, vehicle crashes, carbon monoxide poisoning, and chronic medical conditions complicated by the storm. The actual number of deaths was more than 700, according to BuzzFeed News.[25] The economic loss, as per the state comptroller's office, was between $80-130 billion, but the American Society of Civil Engineers estimated that it could be between $200 and $300 billion.[26]

Texas energy is generated from a variety of sources, but the majority is supplied by natural gas, followed by wind and coal. Many failures, we later learned, were related to the state's supply of natural gas being impacted by the freezing temperatures' effect on the equipment used to produce and transport natural gas, as well as the power at the plants that turn natural gas into electricity. The Lone Star State had similar issues in 2011, but the deficiencies were not fully fixed. Had they been fixed, it is possible the devastation of Winter Storm Uri would have looked different.

Besides being a major producer of oil and natural gas, Texas is one of the top producers of wind and solar energy. Texans would benefit greatly if politicians adopted policies to reduce the state's dependence on fossil fuels. Uri's consequences will reverberate for a long time, reminding us of the physical and mental pain, loss of lives, and major disruption that many Texans went through. I, like many Austin residents, know of families who had to stay out of their homes for months as they waited for repairs to broken pipes that damaged structural frames, wood framing, walls, and ceilings.

OUR CHALLENGE

The above examples of extreme weather events show the severity of the climate crisis, which includes the impacts of heat, severe drought, flooding, wildfires, melting glaciers, and sea level rise. As professionals in the building industry, how do we plan, design, and construct buildings and infrastructure to withstand such forces as wind, rain, and rising temperatures? This monumental task requires all of us to gain knowledge of what is happening, listen to facts explained by scientists, and partner with public and private entities around the globe to find solutions. Scientists, engineers, economists, and researchers tell us that it is possible to stabilize the climate if we start working together with the right policies in place. A tree planted today will bear fruit in the future.

Victimized people who have filed insurance claims are painfully aware of how difficult it is to insure properties in the path of recurrent and severe climate events. This is not to undermine those victims who could barely get by and felt insurance was not one of their needs but shelter and livelihoods were more critical. Often, homeowners of damaged property find out how little coverage they receive for floods, wind, and rain after the fact. Insurance companies are overwhelmed. Many are raising rates and not providing coverage in vulnerable areas. According to the *Texas Tribune*, Texas insurance premiums increased twenty-two percent on average in 2023,[27] twice the national rate. Among the highest number of billion-dollar disasters experienced, Texas leads the U.S. in total cumulative costs (~$402 billion) from billion-dollar disasters since 1980.[28]

In a *60 Minutes* interview on CBS regarding damages from Hurricane Ian, one insurance company executive mentioned that some residents didn't know that wind and rain coverage had to be separately purchased. Imagine that you have lost your property, and your insurance coverage is not adequate to fix your property. Even when help is available from federal, state, and local governments, it takes a long time to access your coverage, and it may not be adequate.

BABCOCK RANCH – A BRIGHT SPOT

The Babcock Ranch was a bright spot north of Fort Myers, Florida, when Hurricane Ian hit on September 28, 2022. In this innovative community, homes were built to withstand forces that Mother Nature could throw at them without suffering floods or a loss of electricity, water, or internet. Babcock Ranch is located thirty miles inland, which is crucial during a hurricane. The town has a field of solar power to generate enough electricity. Power lines are underground to shield them from high winds. Giant, natural retaining ponds surround the development to protect houses from flooding. As a backup, streets are designed to channel floodwaters and spare the houses.[29] This shows that we know how to plan and design our communities if we put the right policies in place, revise building codes, and prioritize design strategies that respond to climate.

WARNINGS EVERYWHERE

U.S. Senator Sheldon Whitehouse of Rhode Island, who chairs the Senate Budget Committee, said at a February 15, 2023, hearing,

"We have all these warnings, warnings of crashes in coastal property values, as rising seas and more powerful storms hit the thirty-year mortgage horizon. Warnings of insurance collapse from more frequent, intense, and unpredictable wildfires. A dangerous interplay between the insurance and mortgage markets is hitting real estate markets across the country. Inflation from decreased agricultural yields, massive infrastructure demand, and trouble in municipal bond markets."

So, what can we do to protect our lives, livelihoods, and properties? We know that we cannot continue burning fossil fuels and spewing harmful emissions and hope that nature will somehow take care of it or that future generations will handle it. A business-as-usual approach will bankrupt us and hurt our health and safety.

When we get sick, many of us trust medical doctors and let them cure us. We get advice from medical professionals when our loved ones are suffering. So, why don't we listen to climate scientists when it comes to global warming? A vast majority of climate scientists agree that the increased use of coal, oil, and natural gas is responsible for increasing polluting emissions and rising global temperatures.

I have a distinct memory of my childhood years in Ahmedabad, India, of sleeping on the roof terrace and waking up to clear, cool mornings. Today, the air in that region feels thick and unhealthy. With the heavy use of fossil fuels, polluting emissions have risen everywhere. Along with an increase in the average humidity level and tailpipe emissions, the air is especially bad for those who suffer from respiratory issues. Health effects vary according to the location, extent of pollution, economic conditions, environmental conditions, and actions by governments of that region.

Looking at India's past has helped me find clues to answer our crisis today. In the next chapter, let's explore how humanity lived in harmony with the surrounding environment. Can we find similar comfort and joy for many years to come? How can we apply the historical lessons they left us? In what innovative ways can we combine the resources and technological skills we have today?

CHAPTER 2
BUILDERS OF THE PAST SHOWING THE WAY FOR THE FUTURE

Historical architecture is fascinating. During my travels around the world, West and East, I have visited buildings and amenities skillfully designed and constructed without any modern technology. During one of my travels in India, I wanted to learn about the Indus Valley (now known as Harappan) civilization and how our ancestors had built cities that thrived 4,500 years ago.

HARAPPAN CIVILIZATION AT DHOLAVIRA

Like most worlds of the past, the Harappan urban civilization was accidentally discovered when early twentieth-century British railroad builders came upon an enormous ruin full of baked bricks near the Ravi River in Punjab. This find in what was at the time British India began a fruitful archaeological journey that led to the discovery of many urban settlements. The grandest was Mohenjo-daro in Sindh, on the banks of the Indus River. Initially called the Indus Valley civilization, the area is now mostly known by its more prosaic name, the Harappan civilization, because of its enormous expanse. Many sites far from the Indus were an integral part of the mature civilization that thrived from about 2600-1900 BCE. Both Harappa and Mohenjo-daro are in what is now Pakistan.

In the middle as well as the latter part of the twentieth century, the Archaeological Survey of India (ASI) embarked on an aggressive program to unearth Harappan sites in post-Partition India. The success was nothing short of spectacular. Major sites were dug from the state of Haryana to Gujarat, including Lothal, Kalibhangan, Rakhi Garhi, and Dholavira, demonstrating the vast area covered by the largest ancient civilization contemporary with Egypt and Mesopotamia.

I recently visited Dholavira, which is the best Harappan site in India according to the UNESCO World Heritage Convention.[1] Our gifted ancestors had built cities that were well-planned. Building construction was done in stone masonry with mud brick cores, brick floors, and wooden superstructures. These cities were among the first known to have prominent infrastructure for stormwater drains, water harvesting, and sanitation devices. The well structures, reservoirs, and baths were built with the same bricks and stones, which held them together for 4,500 years, as evident in the ruins shown in the following pictures.

Professional architects, engineers, and builders would appreciate the lasting achievement of building stone-and-brick walls with a mortar mix of lime, mud, and bitumen that could combat the forces of nature for this long. Local materials and environment were front and center in the planning, design, and construction of these ancient cities. The people were also exchanging skills and techniques with other civilizations with whom they had established trade relations.

The Harappan site of Dholavira is in my home state of Gujarat, India. I am in awe of the grandeur achieved by our ancestors 4,500 years ago. Dholavira is surrounded by terrain of white salt, making it an intriguing location. Why build a city in the middle of a dry desert? Today's salt marshes would have been about thirteen feet underwater when the city was thriving, thus making it a crucial hub for the Harappan civilization's substantial overseas trade with Mesopotamia.[2] Like many Harappan cities, it was

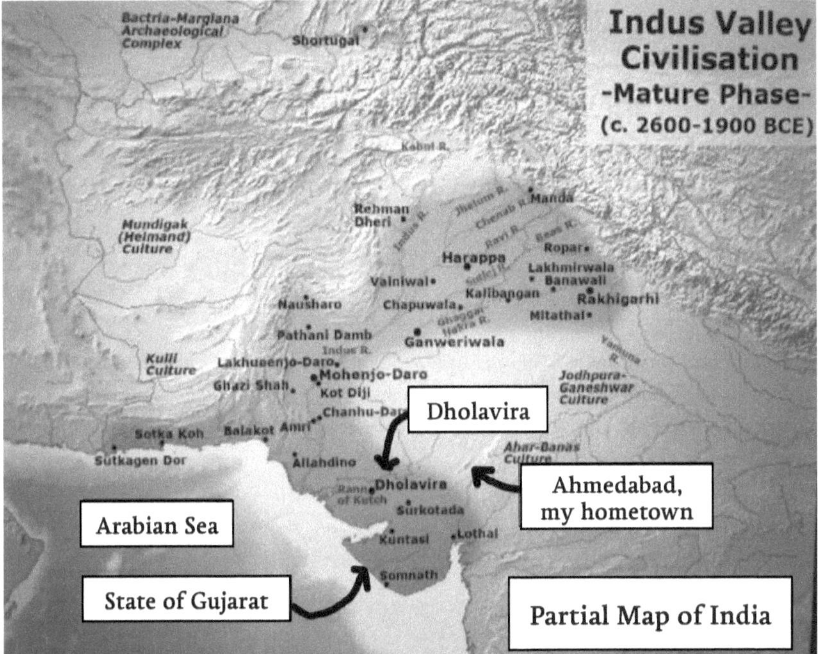

Fig. 1. Indus Valley Civilization (in shaded area) spread from the foothills of the Himalayas southwest towards the Arabian Sea and the state of Gujarat, which is my home state, extending roughly 900 miles. The city of Ahmedabad is my hometown, which is about 270 miles from Dholavira that I visited. Harappa and Mohenjo-daro were the largest discoveries. These sites were part of India prior to the Partition of India and Pakistan. **Dholavira** is considered the best Harappan site in current day India.
Source: https://en.wikipedia.org/wiki/Indus_Valley_Civilisation (2600-1900_BCE).png

planned and built with water as a principal necessary component. Archeologists identify the areas as Lower and Middle Town, with a large open space used as a stadium or a place for public gatherings. The archeologist R.S. Bishat, who headed the Dholavira excavation, described the city as having "excellent town planning with mathematical precision. The entire site was mapped out and divided into squares and triangles, and there was a definite ratio and proportion to each minor and major division."[3]

The experience of walking through the ruins was emotionally overwhelming. Our ancestors were wise to use and apply their knowledge of the surrounding elements of nature

Fig. 2. Large and small water reservoirs used stone walls and stone channels. A stair was constructed to possibly access the drain and manage the reservoir. The reservoirs were strategically located around the city and were connected through stone ducts, linking the two local bodies of water.
Source: Photos from the author's collection

Fig. 3 - **Left.** Stone water duct with carefully selected lintel stone piece.
Fig 3 - **Middle.** Stone drainage trough.
Fig. 3 - **Right.** Stone walls: mortar has held these stone walls together for some 4,500 years.
Source: Photos from the author's collection.

Fig. 4 - **Left and Right.** Current Dholavira village homes, residents with author's husband.
Source: Photos from the author's collection

and local climate. The hot and arid climate had its monsoon season, but during the rest of the year, crucial to their survival, they used an ingenious water management system with strategically located wells and reservoirs. Besides baked bricks, they used sandstone of different colors for the construction of walls of the fort, castle, wells, homes, and sports arena with consideration to strength, aesthetics, and quality of construction.

This humbling visit to Dholavira boosted my confidence! Building long-lasting and good-quality structures is an important part of sustainability. Architectural masterpieces were built all around the world and have survived for years. Other examples that I have visited include the Acropolis in Athens and historic sites in Greece, Italy, India, and Kyoto, Japan. Why not take a mental blueprint from their genius and apply it?

The people of the Harappan civilization eventually moved to the south and east. Harappan's decline could have been due to earthquakes, floods, or droughts.[4] Ancient civilizations and cities have been abandoned due to lack of water, too. One such ancient city was Fatehpur Sikri.

PLANNED CITY OF FATEHPUR SIKRI

India was a prosperous country with a vast empire during the reign of Mogul Emperor Akbar. After winning many battles, he decided to construct a new capital for his empire in 1571 ACE on the hot and arid land near Agra. With his vast war chest, Akbar hired the best architects and artisans to build an imposing city with palaces, buildings, water ponds, and fountains made from red sandstone and marble. The Mogul architecture in India flourished. Architects and builders gathered skilled local craftsmen, including those from outside India. They constructed buildings

Fig. 5. Red sandstone buildings of Fatehpur Sikri.
Source: Miki Desai's photo collection

with large stone overhangs for windows, provided needed shade and allowed airflow without direct heat through doors and walkways, and designed a distinct visual image for the city. The city's stone buildings are still in good shape after being exposed to elements for 450 years. The Taj Mahal, an architectural marvel, was built near this site sixty years later.

Akbar called the city "Fatehpur Sikri" (City of Victory). Yet, fifteen short years after its completion, the Mogul Emperor's capital had exhausted its water supply and had to be abandoned. The groundwater had dried up. Even with Akbar's vast wealth, survival was not possible without a water source. This abandoned marvel, while a colossal failure of planning, remains an impressive ruin visited by many tourists. As I noted in Chapter 1, water supply is a big issue in the foothills of the Himalayan region. Current growth in the capital of New Delhi, which is about 300 miles south of the Himalayan region, has forced residents to dig wells as far as 1,000 feet into the ground. My Delhi family is dependent on such wells.

THE STEPWELL OF ADALAJ

For centuries, bodies of water have been integrated into architecture and social life to combat unbearable heat. In India's dry western states of Gujarat and Rajasthan, the ruling authorities used architects and builders to construct elaborate underground structures called *stepwells*. Along with being a well, a stepwell was a social gathering place for villagers to take shelter during the months when temperatures languished above 100 degrees Fahrenheit for days. A stepwell, true to its name, had steps going down four stories to the well for villagers to fetch water and get respite from the heat. Stepwells provided a place for residents at all levels to socialize.

The Adalaj Stepwell was built in memory of Rana Veer Singh by his wife, Queen Rudadevi, for the common people in 1498 ACE. She held festivals there. This unique work of beautifully designed architecture provided comfort and joy during the hot climate. Its remarkable construction has withstood the thrust of soil from all sides and remained in good shape to this day for more than 500 years.

My undergraduate thesis, called "Response to Climate," at the School of Architecture in Ahmedabad, India, was about a variety of building forms in different types of climates. The form of the Adalaj Stepwell is one example. The research centered on climate conditions in urban and rural areas of hot and arid, temperate, hot and humid, and cold climates, with a specific emphasis on hot and arid climate homes.

VERNACULAR ARCHITECTURE[5]

Throughout history, humanity has been building shelters to find protection from the surrounding environment. *Vernacular Architecture* refers to structures around the world skillfully built

Fig. 6. View from the Adalaj Stepwell's first floor.
Source: Photos from the author's collection

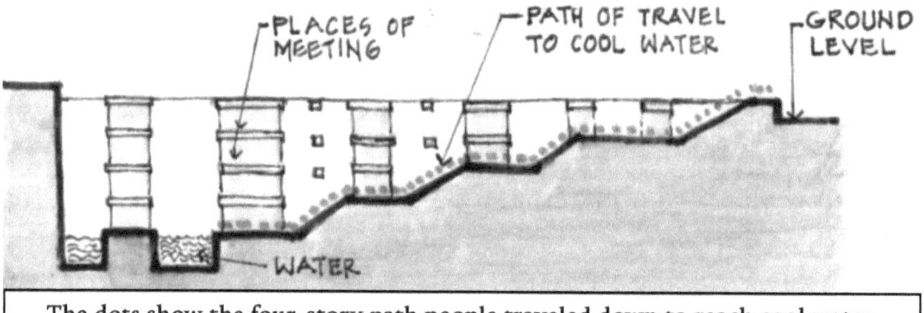

The dots show the four-story path people traveled down to reach cool water.

Fig. 7. Cross section of Adalaj Stepwell.

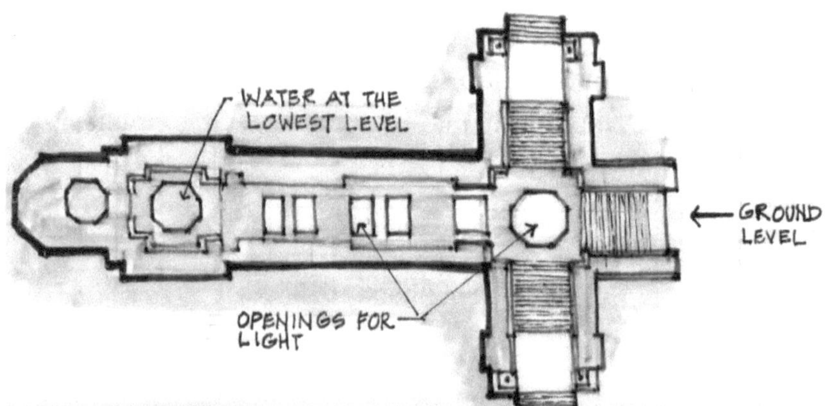

Fig. 8. Plan of Adalaj Stepwell from the ground level down to the water.
Source: Figs. 7-8 from author's undergraduate thesis

by common people without being formally designed by architects or designers, using local materials and construction techniques. They are living examples of climate-responsive buildings.

Communities obviously built structures for thousands of years prior to the Industrial Revolution, when electricity was not available. However, the inhabitants of these communities lived in harmony with the environment. They gained knowledge of the sun, wind, rain, and vegetation. Their architectural forms and lifestyles expressed their local climate. As a result, locals were able to gain maximum comfort, whether in the Sahara Desert or the Arctic.

Some basic principles bind these vernacular structures together despite local variations within each type of climate community. When people laid out their streets and homes, they had to respond to their climate, but they created spaces for the enrichment of life along with physical comfort. The deepest roots of any culture are as immersed in the environment they develop as they are in the attitudes toward that environment.

HOT AND ARID CLIMATE HOMES

Jaisalmer, India; Mesta, Greece; and a village in the state of Punjab in India were examples of hot and arid climates where inhabitants chose dense and tight communities with narrow streets and continuous, high-volume dwellings to maximize shade. High volumes gave varied thermal conditions, with higher levels being hot and the lower levels being the coolest. Winding streets, when oriented properly, moved volumes of air, providing ventilation within the homes. Flat roofs were used for nighttime sleeping since the dry air cooled quickly at night. Homes were small with a few rooms, and inhabitants had multiple uses for all spaces, including the roofs for socializing. Courtyards provided light, fresh air and places for social gatherings, food preparation, and dining. Doors and windows were made of wood, with steel bars for security when they were left open. Outside open spaces were heavily used during mornings, evenings, and good weather days.

COLD CLIMATE HOMES

In the cold climates I studied—Srinagar and Bhrampur in India—homes are generally located near hilly areas toward the northern latitudes. Builders laid out these communities with high-volume dwellings and narrow winding streets that sheltered the warm air at home entrances. Site location was crucial. Finding pockets of land around the mountains and planning carefully oriented streets provided protection from cool breezes. Residents entered and exited using lower levels with small openings in winter. Upper floors had large openings that were used in summer for light and ventilation. Sloping roofs helped during heavy rains and snow. People used outdoor spaces in the summer and during good weather days.

TEMPERATE CLIMATE HOMES

In temperate climates, I studied the Greek island of Hydra and a Japanese village. For most of the year, temperate climates were pleasant, allowing communities to have open spaces with vegetation. Homes had variable heights. Streets had no need for a definite character. The Greek island of Hydra developed in constrained spaces and was tightly packed. But despite space restrictions, it had outdoor spaces incorporated with each dwelling that allowed flowers and vegetation to grow. The materials used included stucco on stone and brick with terraces to enhance outdoor use. Unlike in hot and cold climates, outdoor spaces were extensions of the indoors and used throughout the year. The Japanese village homes were built with local materials like bamboo and wood. One major difference in temperate climate homes was randomly laid out homes and streets around vegetation.

HOT AND HUMID CLIMATE HOMES

For the hot and humid climates, my examples of places were Dang and Byrlutighdem, villages in India. Houses were loosely laid out in a random order to protect from heavy rain and muddy soil. Open spaces were not part of the dwelling for protection from rain. Semi-covered outdoor areas extending from the house were commonly used to enjoy the semi-covered outdoors. The land was heavily covered with small plants as well as large trees. Ventilation inside was the only way to feel comfort, and that was achieved by perforated walls made from surrounding vegetation. Sloping roofs were critical to keep rain out of the interior. As global warming is on the rise, heat and humidity together create a very high heat index, making spaces uncomfortable without fans and air-conditioning.

SUSTAINABLE ARCHITECTURE

Typically, in the past, these climate-responsive shelters were built by the inhabitants themselves. Architects around the world worked with the inhabitants in the design and construction of shelters, providing crucial help. One such architect was Laurie Baker.[6] Born in England, Baker spent forty years of his life in Kerala, India, and became an Indian citizen. He was known for his use of locally available materials, such as brick and mud, to build energy-efficient, low-cost buildings that were sustainable, comfortable, and eco-friendly. Today, he is called "the Gandhi of architecture" for his years of work in technical and vernacular architectural growth through research and development, educational programs, and rural housing.

Some architects, including colleagues of mine from the School of Architecture in Ahmedabad and in the office of Abhikram, have focused their work on vernacular architecture, conservation, and environmentally responsive designs while restoring local skills and construction methods. My friends Madhavi and Miki Desai have studied and documented vernacular architecture in India for many years.

Miki Desai's research and documentation resulted in a vast visual and drawn repository that he published in the remarkable book *Wooden Architecture of Kerala.*[7] These temples, palaces, and dwellings built in a unique system of wooden construction are 200-600 years old. In the wet climate of Kerala, skilled craftsmen and artisans developed wood construction practices using traditional knowledge systems. In most house types, a courtyard is a central space that acts as a symbolic fulcrum, or as a climatic device with cultural connotations. The form of architectural elements-oriented climatic controls is important to this genre, irrespective of their economic class. As Miki describes, "This is much evident in the coherent and sustainable synthesis of geology, climate, building materials, and lifestyle that culminate in architectural expression."

PLANNING OUR ENVIRONMENT WITH CARE

Vernacular architecture and traditional communities have lessons for us. We can relearn how to work with the surroundings: the site, the landscape, wind direction, rain pattern, orientation, and location of buildings. We can use locally available materials to construct comfortable spaces for inhabitants.

Yes, today's environmental issues are complex due to all our activities over many years. We have incorporated technological devices for comfort and joy, which were possible due to energy sources like coal, oil, and natural gas. We have spewed many gigatons of carbon dioxide (CO_2) and other pollutants into the atmosphere, and we continue to emit more. Our land, environment, and oceans are significantly impacted by changes in climate. Large-scale population migrations occur when residents suffer from a lack of food and water. People need either to sustain themselves or move out of affected areas. Everyone keeps asking what we can do about it. The first thing we can do is learn from our not-so-distant past.

As I shared in Chapter 1, communities continue to lose lives and livelihoods because of extreme weather events. My interest in the environment began as a child growing up in my grandfather's house in Ahmedabad, India. We lived in a hot and arid climate with few resources that offered comfort from the heat. Although we had electricity, its availability and use were minimal. It was critical to build shelters using passive climate elements to get protection from heat or cold. Air-conditioning was not reachable, even in our thoughts. So, how did we find joy and comfort in my grandfather's house?

CHAPTER 3
CHILDHOOD HOME IN INDIA

DISCOVERING MY GRANDFATHER

My grandfather built a two-story white home in Ahmedabad, India, where electricity was available in limited quantities in the early 1930s. My father was a young child at the time. Many years later, my three brothers and I grew up in this house in the 1950s and 1960s.

We called my grandfather *Bapaji* and my grandmother *Ba*. Bapaji was a strict, disciplined patriarch of the family and had his own way of conducting household functions. He made all the decisions, and all of us, as his grandchildren, were afraid of him. We never questioned him. But behind this stern personality, he was a concerned human being who cared deeply for his family and environment. I understood the value of saving resources he had inculcated in us only later, when I started inquiring of family members and some relatives.

Bapaji was born in British-occupied India in the late nineteenth century, prior to the start of Mahatma Gandhi's freedom movement, which originated in our hometown of Ahmedabad. Bapaji raised us as strict Jains,[1] one of the religions in India practiced by less than one percent of India's population today. Before I moved to Austin, Texas, I spent my entire childhood and college years in my grandfather's house, along with my siblings, parents, and my aunt. It was a large patriarchal family, and we all have special memories of climbing trees and playing games outdoors on good weather days and indoors during the hot summer season. In a hot climate, our family had a practice of gathering outside or taking walks early in the morning or after sunset. Outdoor spaces like balconies, verandas, courtyards, and terraces were an integral part of this house.

This home still has a marker on it saying *Sutaria Niwas*, house of the Sutaria family.

My childhood home at 38 Jain Society in Ahmedabad is ninety years old now, and my cousins still live there. The perforated balcony and verandas (draped here) open from both sides and are the most popular places in the house as air flow is crucial for comfort. The exterior is painted with whitewash, a combination of slaked lime and calcium carbonate, which reflects the sun and reduces heat inside the home. The flat roofs are used for sleeping in summers. Its other uses include kite-flying celebrations and drying of foods and clothes.

Fig. 1. Author's childhood home in Ahmedabad.
Source: Photo from the author's collection

MOVING FROM A HISTORIC CITY

Bapaji grew up in a middle-class family and received an elementary school education in Ahmedabad. He had few resources to start his life, but he had a strong will to build a bungalow and create a better life for his family. The fort city of Ahmedabad is 612 years old, and its historic homes are believed to be 300-400 years old, giving Ahmedabad the unique distinction of being a UNESCO World Heritage City.[2] Buildings are organized in tightly packed communities called *pols*,[3] and residents have strong social ties within each pol.

Homes were built to combat the unbearably high summer temperatures in Ahmedabad, and every design feature reflects some response to the city's harsh climate. The features that create each pol's unique aesthetic character include narrow shaded streets, common walls, interior courtyards, ornamental facades and doors, windows, and balconies with patterned grills and overhangs.

The courtyards inside these homes were used for multiple purposes. They allowed light and air to also be used as social gathering spaces. With enough light and air from the courtyard, residents could use any of their balconies based on the time of the day as well as the season. Many courtyards once had water storage tanks underground, which were used until the city provided municipal water connections.

These homes worked well when a single family owned their entire residence. Bapaji didn't have that luxury. My granduncle had a small home in one of the pols. Bapaji moved in with my granduncle after leaving a small town for opportunities in the large city of Ahmedabad. Bapaji and Ba continued to live there with their two children, but the place was overcrowded with two families. He essentially had one room for his own family while they shared the kitchen and washroom facilities. A major drawback of the pol homes was the lack of light and open land for plants and trees. Once you get out of the pol, the narrow streets open wider to plazas, temples, and small parks.

Bapaji had heard about newer neighborhoods on the western side of the river Sabarmati, where people were building Western-style bungalows on large open lots. He walked to the newer parts, and, after careful thought (in true Jain fashion), he decided to purchase land near one of his Jain relatives who had moved there. Despite the hot, arid climate, this part of town had large *neem*, *ashoka*, and fruit trees. They provided shade, and locals used their leaves and fruits for homemade medicines.

BUILDING A NEW HOME

Ahmedabad had a thriving textile industry, which offered opportunities to cloth merchants for wholesale and retail businesses. Bapaji worked in one of the retail shops and started saving. Being a strict Jain, he first purchased a plot in a community called Jain Society, where a Jain temple was already built within a short walking distance. Visiting a temple every day was important to him and Ba.

The plot Bapaji purchased was a large lot with many beautiful trees. The new construction was influenced by British colonial architecture mixed with vernacular designs. The developer for this community had one floor plan for two-story homes, and he modified each plot to save the property's large trees. Bapaji managed to construct the entire two-story home with savings from his modest income. To this day, I remain impressed by his determination and resourcefulness to plan and build a home.

He built the house over the course of several years, room by room, starting with a kitchen and then another room. The bedrooms and living room were on the ground floor. With limited funds, the upper floor was the last part of the house he finished. Once completed, the house had fourteen-inch-thick brick walls with lime plaster and whitewash to keep out the oppressive heat. In the 1930s, cement was a commodity that was not available in large quantities. To save money, a common practice was to use lime in mortar and plaster to reduce cement content in the construction.

Today, professionals in the building industry know that cement has a large carbon footprint. It contributes to warming the planet. In fact, global research efforts are ongoing to reduce cement content in construction. What are we replacing it with? It is not a surprise that thousands of years ago, Indian and other craftsmen in the world knew how to build masonry walls with lime. Researchers continue to learn from the past to find solutions for the future.

Fig. 2. Mosaic tile roofs in the old city of Ahmedabad. They are effective in keeping the heat out.
Source: Photo from the author's collection

Fig. 3 - Left. Swinging on the veranda is still a favorite place of the author. The veranda is covered but open on both sides. "When I used to visit this house, I enjoyed watching the birds, including the occasional appearance of peacocks."
Fig. 3 - Right. The living room doors with grills; these doors are still functional after ninety years.
Source: Photos from the author's collection

Ahmedabad had long summers, with temperatures rising above 105 degrees Fahrenheit for months prior to monsoon season in the 1960s and 1970s. Roofs provided openness and cool breeze, and they had to be constructed using steel and concrete. Due to large variations in temperatures, these flat concrete roofs would also develop cracks, creating water penetration problems. Leaky roofs with buckets below to collect water were a pretty common sight during monsoon season. To fix the cracks, our roof was a patchwork of asphalt paste, crisscrossing the entire roof. Wealthy people had roofs covered in beautiful floral designs out of broken pieces of mosaic tiles called "*kapchi*," which helped to seal the concrete surface and stopped water penetration. They used white mosaic, for the most part, to help their roofs reflect heat and lower inside temperatures.

Bapaji couldn't afford to put mosaic tile flooring on our roof. I know from my current work that roof repair projects do come up when high-quality roofs are not built. Why are they so important? Water penetration degrades materials, lowers the longevity of steel and reinforcement, adds humidity, and increases maintenance costs. A good-quality roof is also the most sustainable construction.

As a child, being on the roof after sunset was a cool experience in every sense of the word. Ahmedabad's air was not polluted, and the skies were clear. I fondly remember watching the stars at night and the feeling of a cool breeze touching my skin in the early morning hours prior to sunrise, a truly satisfying feeling. During monsoon season, sudden nighttime rains would send us running down the stairs with our mattresses and pillows. Indoor sleeping was uncomfortable with high humidity and no fans! Windows and doors with grilles were open to allow ventilation.

The majority of the home's wooden doors and windows are still in reasonable shape after ninety years. Balcony railings were designed both for aesthetics as well as to capture the breeze.

Glass was expensive, so its use was minimal. Small panes of wire mesh glass in the windows provided daylight. Swings on verandas were and are still popular.

Ba and Bapaji's entire family of fourteen—including Bapaji's eldest son from his first marriage and my father's family—lived on the second floor. He had rented out the ground floor because he needed the income to complete the full construction, support the family, and educate all his grandchildren. Everyone used one staircase outside to go up and down. The water closet was separated from the main building in an outdoor building with an Indian-style toilet, with a hole in the floor. The building was raised about four feet above ground to allow workers to clean the waste from below, which was picked up by the municipal corporation's sewage trucks. The house had no sanitary sewer lines at that time, except for the drains in the kitchen, bathroom, and bedroom.

Eventually, the tenants vacated the ground floor. My uncle and his family stayed upstairs, but the rest of us moved downstairs. However difficult life was, we could not express enough joy to have more space and cooler temperatures on the ground floor—and a connection to the land! We had municipal water service for a couple of hours each morning and evening. Water had to be stored in buckets and large containers. The municipal corporation also provided electric service, but Bapaji wanted minimum use of it. We had a few incandescent light bulbs and plugs in every room.

My father, like me, had a hard time dealing with the heat. He asked for ceiling fans, but Bapaji didn't approve. It was unbearably hot when outside temperatures were 100 to 105 degrees for months in the summer before monsoon season. Occasionally, I used to get heat strokes. In May of 2024, the heat in India was oppressive, with off-the-chart temperatures causing school kids to get heat strokes in classrooms without fans. Many daytime activities of ours took place in an open area called the *chowk*. It was lined with mattresses for sleeping at night. Sleeping inside was torturous. In summer, especially during the monsoon season, we also suffered from various diseases, such as malaria, and endured other health issues. The monsoon brought relief after months of heat stress, and that emotional relief is often expressed in Indian folk songs, dances, and films. Indian classical music devotes certain melodies to monsoon joys.

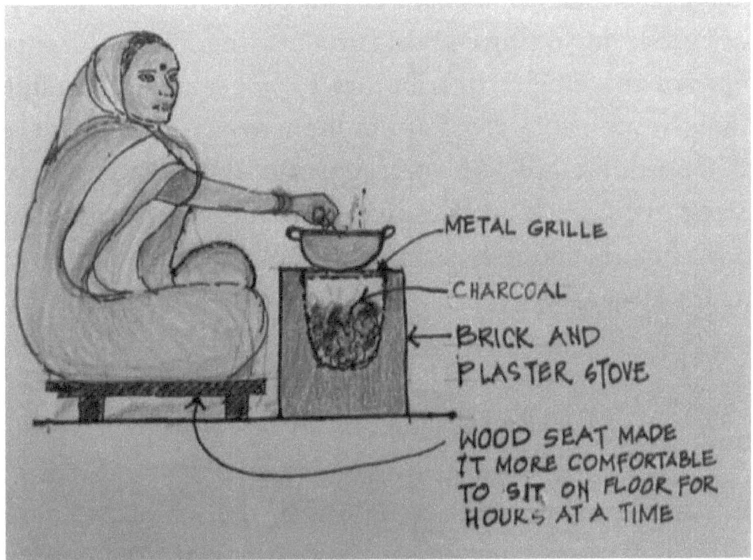

Fig. 4. In Bapaji's home, a typical cooking method with a coal stove included daily use of charcoal. A wooden seat raised the level and added comfort to sitting on the floor for hours.
Source: Author's sketch of her mother

Finding comfort from the heat has been my passion ever since growing up in this house. Our mothers sat on the floor and cooked our meals using coal and kerosene stoves. Fortunately, we had windows and cross ventilation in our kitchen. In his book *Drawdown*, Paul Hawken reports that forty percent of humanity still uses these cooking materials, often in unventilated spaces.[4] This is a serious health concern in remote villages, but the problem requires attention from governments as well as international organizations to devote resources to better stoves. These villagers need fuel to feed their families. Their resources are minimal, but better stoves will improve their health and be better for the environment.

Fuelwood remains the principal energy service provider to seventy percent of the Indian population.[5] This means that roughly 900 million people are exposed to air pollution daily in India. Globally, three billion people utilize biomass fuels such as wood, charcoal, animal dung, and plant residues for everyday activities, including cooking and heating. The resultant pollutants are the major contributors to household air pollution, which is responsible for four million deaths worldwide each year.[6] Environmental and some government organizations are investing in clean-stove initiatives to reduce emissions, improve women's health, and reduce deforestation. We need more investments for larger impacts.

RELIGIOUS PRACTICES OF MY JAIN FAMILY

Jainism is one of the oldest religions in the world and dates to 3,000 BCE. Jainism is based on the concept of *ahimsa* (nonviolence in Hindi). Jains believe all living beings, no matter how small, have a soul and should not be harmed. Jain *sadhus* (religious leaders) wear a mask-like cloth over their mouths to avoid swallowing insects.

Jains are stricter vegetarians than other Indian vegetarians. Real believers of Jainism do not eat anything that is grown underground. That includes onions, garlic, ginger, potatoes, and other root vegetables like carrots, radishes, and beets. They believe that harvesting them would not only kill many living beings in the ground but also that these vegetables are covered with insects and could have adverse health effects.

Bapaji and Ba were real believers, and they raised us as strict

Jain vegetarians. I ate onions, potatoes, and other root vegetables only after I left for college. Dairy products are acceptable in Jainism, just as in Hinduism, and milk products are used in homes and temples. In today's Jain households, young people are giving up these strict vegetarian practices. Jains are totally obsessed with food and spend a lot of time on food conversations and preparation.

REDUCTION OF WASTE

We always ate at home except when we were traveling to nearby Jain temples. Cooking consumed most of a woman's day. Bapaji strictly followed Jain religious practices when selecting food and leftovers. We were not allowed to eat leftover cooked food the next day unless they were fried things or dried crisp bread. We didn't have a refrigerator or a cold box to store cooked food. This practice was known to those who worked inside and outside of our homes. Workers would come by every evening to col-

Fig. 5. Vendors on the street sell clay pots for water storage. The practice of using a clay pot for drinking water is common among all types of families in Ahmedabad.
Source: Photo from the author's collection

lect the leftovers, or children were sent to convey a message for leftover food pickup. Food was not wasted. Raw vegetable scraps were fed to stray animals. We also saved every scrap of paper, cloth, and metal to sell to textile, paper, and other industry recyclers, who would come collect the scraps. Waste has increased in our homes, but the practice of recycling continues today in small and big towns.

We stored our drinking water in clay pots to keep it cool, which generated minimum landfill waste. This practice of storing water in thick clay pots is still prevalent in India and used in many homes. In fact, clay cups are now introduced by some vendors for tea and coffee to reduce the use of single-use plastic, which is banned by the Indian authorities.

Even today, the leftover cooked food in the homes of my family in India is given to those who work in the house or others in need. But the total amount of waste has increased in Ahmedabad and in Delhi households due to an increase in the convenient packaging of food. In our Delhi home, everything is thrown into the trash, including food scraps, paper, and plastic, which is taken to landfills. NASA satellites have detected heavy methane emissions from huge landfill dumps in Delhi. Workers who live near these dumps and sort the waste are exposed daily to harmful emissions that affect their health. Although the problem of waste is particularly bad in Delhi, it is prevalent throughout India, even in the remote and beautiful foothills of the Himalayas and on the streets near holy temples.

Wasteful habits take time to break. In Austin's Bridge Center, where we play bridge, the players for every game use a convention card, which is paper that can and should be recycled. Despite the placement of big signs on the blue recycling bins, some players routinely throw the cards in the trash. Reduction of waste is one easy step that we can all practice to cut down on harmful methane emissions, which are rising.

In Austin, we have three different waste containers, including one each for landfill trash, compost, and recyclables. Not all single-family homeowners participate in composting, which can reduce waste. City officials have made efforts, but they must find ways to reach out to citizens and persuade them to compost. The city council recently approved a resolution that requires multistory apartments to prepare for composting for its residents by October 2024, which is a major step toward zero-waste goals. Composting and recycling have helped my household in the reduction and diversion of landfill waste. Food loss and waste are estimated to account for roughly one-third of the food intended for human consumption in the United States. Production, transportation, and handling of food generate significant carbon dioxide (CO_2) emissions, and when food ends up in landfills, it generates methane, an even more potent greenhouse gas.[7]

According to Hawken in *Drawdown*, the food we waste contributes to roughly eight percent of the total anthropogenic greenhouse gas emissions. He notes that "people who need food are not getting it, and the food that is not getting consumed is heating up the planet."[8]

OUR PLASTIC PROBLEM

I hardly remember any packaged foods during my childhood. Lentil varieties, wheat, rice, vegetables, and combinations of various flours cooked in a variety of ways made up our primary staples. Every neighborhood had hawkers who went around selling vegetables and fruits, which customers purchased and brought home in their own containers. Bapaji walked every day with his cloth bags to the wholesale market in the old city and bought seasonal vegetables. Plastic grocery bags were nonexistent. The milkman came on a bike every morning to deliver milk. He would pour it into one of our cooking pans, and my mother would boil it in that pan before using it. Now, the milk is delivered in plastic bags. Like many countries, India today has dramatically increased its use of plastic bags and containers.

In March of 2022, at the fifth session of the UN Environmental Assembly (UNEA-5.2), a historic resolution was adopted to develop an international legally binding instrument on plastic pollution, including in the marine environment. It requested the UN Environmental Program to convene an Intergovernmental Negotiating Committee (INC) to develop "the instrument," which is to be based on a comprehensive approach that addresses the full life cycle of plastic, including its production, design, and disposal. Plastic production has risen exponentially in recent decades and now amounts to some 400 million tons per year—a figure set to double by 2040.[9] Plastic pollution, which contributes to climate change, threatens food safety, quality of human health, and coastal tourism.

Awareness of the dangers of plastic waste has prompted many people to switch over to reusable bags, especially cotton bags, and reusable bottles. But if we look closely, plastic packaging is literally part of everything we purchase. Most restaurants still use polystyrene foam (better known as Dow Chemical's trademarked brand, Styrofoam) containers for take-home food. According to Let's Talk Science,[10] "Polystyrene has many different uses. It can be a great help to us in many ways. But we must reduce our use of it, reuse it, or even recycle it. That way, we'll help prevent it from damaging or polluting our environment." It is expensive and difficult to recycle polystyrene waste, and therefore, doing so is uncommon. I have started carrying my own glass containers to restaurants to avoid taking my leftovers home in Styrofoam containers.

Without influencing the markets and without innovations and individual awareness, it would be hard to reduce plastic waste or its damaging impacts on the environment. Plastic bags and other plastic packaging materials are recyclable at some grocery stores, but the extent of recycling is minimal, according to Bloomberg News.[11]

CONSERVATION OF RESOURCES

Bapaji was obsessed with the conservation of electricity and the Jain practice of nonviolence. He refused to install ceiling fans partly to save electricity use but also because he believed fans killed insects. It was extremely uncomfortable in monsoons when humidity and temperatures were high, and we were forced to sleep inside without ceiling fans. Eventually, my father, as the next patriarch of our family, had ceiling fans installed in the house. They gave us a distinct feeling of comfort; however, many families in hot regions of the world, including poor families in the U.S., cannot afford fans.

In Ahmedabad, we had six to eight weeks of winter with low temperatures between 50-60 degrees and highs between 75-85 degrees Fahrenheit. Although homes were not heated, winter was the most enjoyable time of the year. With dry and clear weather, we would typically have a thirty-to-forty-degree difference in day and nighttime temperatures. *Bumbo* (a bronze water tank) was used for warming water. A small amount of

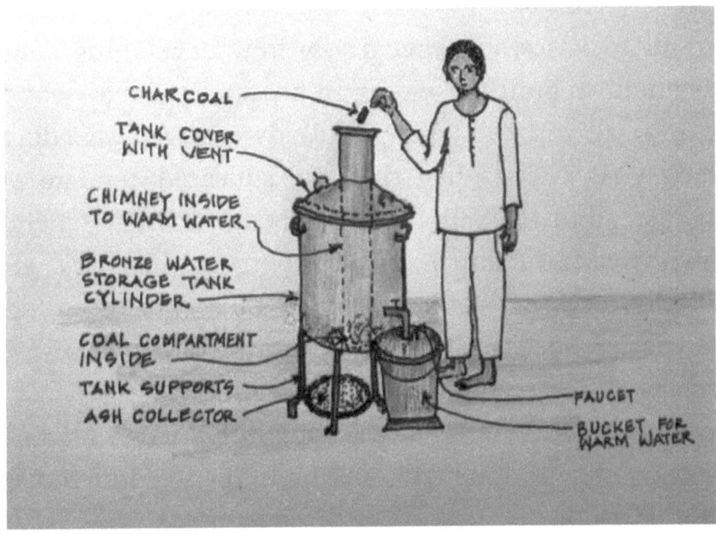

Fig. 6. Using *Bumbo* to warm water in winter.
Source: Author's sketch of her brother

warm water was mixed with cold water to make a one-half bucket, which was enough for a bath. The municipal water we received for a few hours each morning and evening required water storage.

Due to the limited number of clothes, limited storage space, and dusty conditions, washing was a major daily activity. All these activities took place in our chowk, which was open to the sky and had a veranda on two sides, with a perforated railing. Clotheslines for drying clothes were a part of building design. These multipurpose spaces played an important role in homes where large extended families had to figure out how to create comfort in different seasons, including hosting social get-to-gethers, playing games, and sleeping outdoors in the summer. As I shared in Chapter 2, such outdoor and semi-outdoor spaces have played an important role in vernacular architecture around the world.

IMPORTANCE OF EDUCATION

In *Drawdown*, Hawken shows the importance of the education of girls in all societies, but specifically how developing countries would benefit in multiple ways.[12] In my family, Bapaji was determined that all his kids and grandkids received an education, including girls. Girls, when they are educated, tend to pursue careers instead of marrying at an early age and having children. It helps society in many ways, including productivity, growth, and development of women. Women take care of children in most societies, specifically in developing countries, and inculcate values of preserving the environment.

Mahatma Gandhi also had a strong influence in Ahmedabad, and he encouraged women to join him in the freedom movement of India. His *ashram*[13] (hermitage) on Ahmedabad's Sabarmati River is a destination for people around the world.

As a result of this educational encouragement, women got involved in politics, medicine, business, and teaching. I received an education in a joint school for boys and girls. My mother studied in an all-girls school. She was a bright student and wanted to pursue further studies to become a medical doctor. But in 1942, the Quit India Movement[14] against the British took off all over the country. Protests around the country put pressure on the British to leave India. Schools in Ahmedabad had to close for some time.

My father was an active participant in the freedom movement and had to quit school and literally hide himself from the authorities. My mother was already engaged to marry him. At that time, further study for women after marriage would have been unthinkable. My parents couldn't go to college. However, like my grandfather, they, too, wanted my brothers and me to get a higher education, and they fully supported us in doing so.

Bapaji walked to most places and took care of many errands. He rarely took a bus or a *rickshaw* (three-wheel vehicle). The municipal corporation had a public bus system that was the primary mode of transportation for most working people, making them crowded during work hours. Rickshaws were privately run

Fig. 7. Bapaji and Ba with author's father and aunt at a religious ceremony.
Source: Photo from author's family archives

and more expensive. Bapaji normally dressed in a white *khadi kurta* (long shirt) and cotton *dhoti* (wrap-around pants) with a white *topi* (Indian cap) to cope with the heat. Khadi[15] is a hand-spun cloth that was promoted by Mahatma Gandhi. White cotton was produced in the textile mills of Ahmedabad and was the most commonly used fabric. Clothes made from khadi and other cotton fabrics, especially white-color fabrics, are most effective in coping with heat. Ba and other women dressed in a thin cotton *sari*[16] that helped with ventilation around the body in hot climates.

Jains fast routinely on various occasions, either taking a full- or half-day fast or a limited intake. They keep track of the lunar calendar and observe events per Jain rituals. For example, they keep a calendar with exact times for sunrise and sunset. Strict Jains don't consume any food before sunrise or after sunset. Bapaji lived his whole life per Jain practices, was as thin as a stick, and remained in good health until his death at the age of seventy-one.

For most of the year, heat was one of the major determinants for food and transportation choices, cultural events, and necessary social activities. The stresses of life, along with the brutal heat, took a toll on him. On the day he died, I was on the veranda, preparing for exams, when he returned from a walk. He suddenly fell flat after removing just one shoe and passed away. It was a shock to our entire family when the patriarch of the family, someone who was a central figure in our lives, left the world unexpectedly. My father, who was not quite prepared for this moment, took the helm of the extended family. He was committed to educating all children. My father had a great sense of humor and a natural ability to take risks, which helped him through the process of becoming head of the family. We all chipped in taking care of day-to-day chores and were encouraged to educate ourselves through our school years. Without knowing anything about architecture, I wanted to study architecture.

CHAPTER 4

STEPPING INTO THE WORLD OF ARCHITECTURE

SCHOOL OF ARCHITECTURE AT AHMEDABAD

In 1962, the world-famous architect Balkrishna V. Doshi,[1] recipient of many prestigious awards, founded the School of Architecture in Ahmedabad and designed the school's first building. His passive climate design was like a living laboratory, a first lesson for me in modern climate-responsive buildings and the role of architects. In Ahmedabad's hot and arid climate, the building's orientation and location on site were key decisions Doshi made to keep the heat out of interior spaces for a building without air-conditioning.

Even today, windows on the north side are the most preferable because they bring light without direct sun. Professor Doshi designed clerestory windows on the north to bring an abundant amount of cool light and large wood doors with small amounts of glass on the south that can be opened for ventilation. Openings on the south side can be shaded by devices according to the sun's angles. The openings bring cool, indirect sunlight into the space and allow for a breeze. The large wooden doors on the south not only opened to the breeze, but also provided connections from the upper level to the lower floors for social interactions.

The middle level was set back to create a two-story-high outdoor shaded area on the ground floor, which had multiple uses like presentations, social gatherings, lunch places, and other informal interactions. In developing countries like India, the use of passive climate systems was not only a necessity for comfort but also a need for the preservation of electricity.

In contrast, massive brick walls on the east and west blocked the harsh summer sun. Designers avoided openings on the east and west, particularly the west end, as the angle of the sun's rays hit openings on both ends at the hottest part of the day.

In design studios, we prepared drawings by hand, at times without artificial lights. The use of electricity was for lights at night and for fans and plug points. It was not always comfortable during hot days when we had to draw on tracing paper with sweaty hands.

Fig. 1 - Left. School of Architecture in Ahmedabad, designed by Balkrishna V. Doshi. Its north facade with clerestory windows provided cool light in the design studios; author and classmates gathered on the front lawn for discussions during good weather days.
Fig. 1 - Right. South facade of the school with large doors that provide cross ventilation and social interactions.
Photo credits: Vidu Chavda and Dinesh Mehta, the author's classmates

The exposed brick walls and concrete structure are covered with mold in some areas, as shown in Figure 1, and that was a consideration that should have been a part of this beautiful building. Despite the hot and arid climate, the monsoon rains are substantial enough for water to stay on those surfaces, causing unhealthy mold. Water penetration deteriorates building materials, and they must be part of sustainable buildings' design strategy in all climates.

In addition to the design studios each semester, we studied the functional aspects of buildings in subjects like planning, building technology, science, math, climatology, structural design, history of architecture, drawing and painting, and other electives. At that time, we were proud of the breadth of topics included in the curriculum. Only later did I realize that our education didn't quite prepare us for the role of topics like construction technology, landscape design, environmental planning, and the history of Indian and world architecture. The school arranged measured drawing trips to historical places like Jaisalmer, Khajuraho, and Fatehpur Sikri and these trips were enormously informative and educational for lessons from historical architecture.

Our faculty members were trained in the modernist tradition started by the American architect Louis Sullivan, who articulated the principle of "form follows function,"[2] in which the form of a building, or an object, is a starting point for designers and architects to derive the basis of a building's design from the function that the building will serve. Modernist tradition is the architectural style that dominated the Western world between the 1930s and 1960s, and it was defined by a functional approach to building design. Doshi had worked in the offices of prominent architects like Le Corbusier in Paris and Louis Khan in the U.S., who were well-known figures in this tradition. India's first Prime Minister, Jawaharlal Nehru, commissioned Le Corbusier to design the new capital, called Chandigarh, in the state of Punjab, India, in 1952. Corbusier had designed homes and apartments in Spain and France. His designs considered the sun's angles, light sources, and air currents to maximize comfort for inhabitants. They were designed prior to the advent of air-conditioning.[3] He implemented climate-responsive design strategies in housing and public facilities in Chandigarh.

FIRST PRACTICAL EXPERIENCE IN ARCHITECTURE

Our school's curriculum gave us an opportunity to get practical training that helped us with real-world projects and issues. My training at the office of Kamal Mangaldas helped in the preparation of working drawings (construction documents) after the firm's principals determined the initial design ideas. This practical office work allowed us to understand the architects' role in the building industry and in society at large. To this day, I believe that architects can create a better quality of life through sustainable buildings that would benefit humanity, not just focus on pretty designs. Brick, concrete, and steel were major modern building materials at the time. Cement was expensive and in short supply. So, guess what? Lime was used as an alternative. Electricity was expensive; even the wealthy were careful in running their air-conditioning units.

Fig. 2 - Left. Front facade of the Mangaldas home in the old city with shading devices and openings for air circulation. These homes are more than 300 years old.
Fig. 2 - Right. Another home in the same *pol* with balconies and ornamental railings.
Source: Photos from the author's collection

The Mangaldas family had moved to the suburbs of Ahmedabad from the pols. Their beautiful old historic home is now being maintained by his family as a museum. Visitors can see the historic buildings of the pols. The homes (see Fig. 2) have a concrete floor slab within a mainly wooden structure that includes wood windows, steel grills, wood and steel railings, overhangs, verandas, and terraces. Plastered walls and painted wood provided protection from the elements.

UNDERGRADUATE THESIS – RESPONSE TO CLIMATE

My undergraduate thesis was a study of two types of houses: first, the dense and tightly packed homes of the old city pols, consisting of 300–400-year-old homes built prior to the Industrial Revolution, and second, the scattered houses of the new developments built in the early 1970s, specifically homes of middle-class families. During my visits to my maternal grandparents, I had experienced the old homes. The most striking features were narrow streets and cool interiors. The biggest issues were natural light and tight spaces, which drove my family members to move to the suburbs.

The old homes in the pols with their three-story facades had narrow streets that formed interesting pathways. Walking was a major mode of getting around within and through nearby pols. Two major arteries went through the fort city, which offered transportation by bus and rickshaw. Each pol had an open plaza-like area, each with a structure for feeding birds. Residences typically shared common side walls. Fear of invaders was real, so each community had a gate that was locked for security after dark.

The layout of the community and homes that I studied for my thesis (Figs. 5 and 6) are narrow, three-story row houses, each with a courtyard in the middle for light and ventilation. The courtyard was a cool and pleasant place to gather in summer and

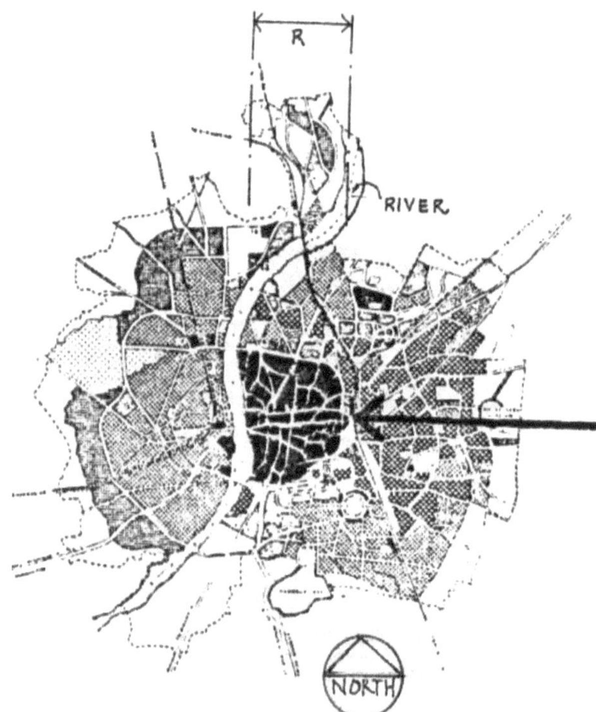

Fig. 3. Plan of the City of Ahmedabad.
Source: Census of India, 1971
Ahmedabad Latitude 23.0225 degrees North. Many pols are within the fort of this old city, shown in black. The horizontal distance from the Sabarmati River (shown in the center) to the fort boundary is roughly 1.5 to 2.0 miles, shown as R in the plan. The homes I studied are located in this blackened area.

Fig. 4. Bird's-eye view of the density within the old city of Ahmedabad.
Source: Photo from the author's undergraduate thesis

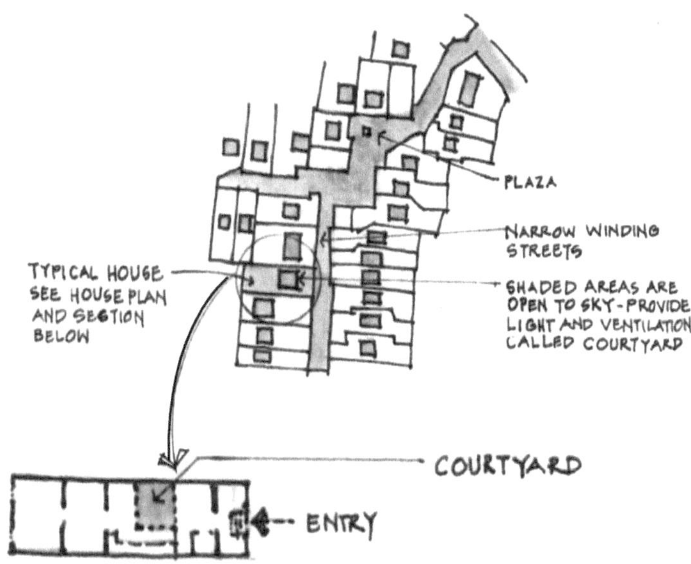

Fig. 5. The layout of a pol and typical house I studied. The courtyard was the center of all activities throughout the day.

Source for Figs. 5-8: "Response to Climate," author's undergraduate thesis

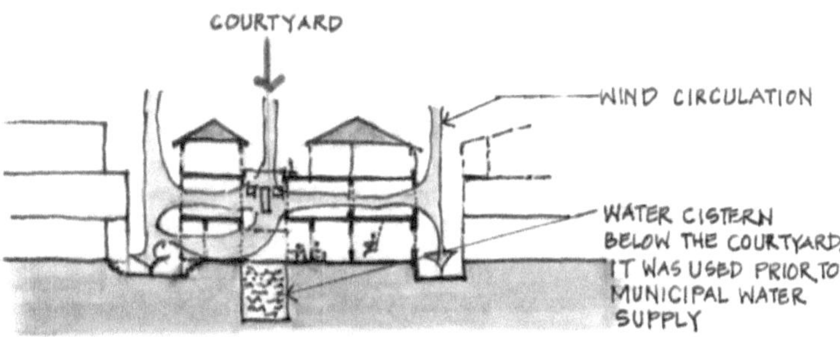

Fig. 6. Cross-section of a home, showing how wind circulation helped with thermal comfort.

Fig. 7 - Left. A view of the courtyard; windows and doors opened inside and on the outer walls, providing cross ventilation, that helped with all activities like for cooking, dining, napping, and socializing.

Fig. 7 - Right. The tile floors felt cool, and it was common to lie down on the floors.

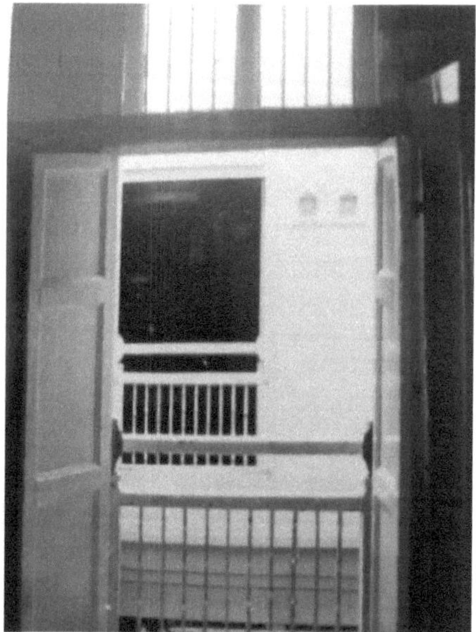

Fig. 8. Railing and doors open to the courtyard or to the street and allow ventilation at a floor level, where activities take place.

became a center of social activities (Fig. 7). The courtyard space was used in multiple ways, including cooking, dining, socializing, and napping. Large extended families occupied these homes, and as families grew, they moved out to newer parts of the city, depending on their economic conditions. These old pol homes provided better protection from climatic conditions, but they had issues accommodating modern facilities, including toilets. For older people, the narrow staircases were difficult to maneuver several times a day. They were noisy and lacked visual privacy since the doors and windows had to remain open for ventilation. The upper levels provided a terrace space to dry clothes and grow vegetables, plants, and flowers.

Fig. 9. Narrow streets of pols provide shade and are still full of activities.
Source: Photos from the author's collection

NEW HOUSING DEVELOPMENTS IN AHMEDABAD

By and large, the planning and design of the new housing developments for middle-class families showed little consideration for the design of dwelling elements in response to climate. Ceiling or portable fans helped to some degree to cope with heat since window air-conditioning was unaffordable to most middle-class families. Having a refrigerator was also a luxury.

Temperature measurements in both old and new homes showed the new homes were distinctly warmer. During the hot days of summer, even a slightly lower temperature made a big difference in one's physical comfort and general state of mind. Heat stroke, diarrhea, and fatigue were the side effects of higher temperatures that many suffered, including myself. Simple design elements could have greatly improved the quality of life for most people: orientation, cool northern light, higher insulation for walls and roofs, reflective and light-colored roof and wall surfaces, well-planned cross ventilation, strategically planted trees and landscaping, and multi-use spaces for various seasons. Such passive climate systems have a large impact and should become part of the building code requirements in India.

Working in the Mangaldas' office, we incorporated such features in the low-cost housing project for the Gujarat Housing Board. However, unless major building code improvements are made and then implemented, many housing projects will not get the benefits of good design and construction. Even when building codes are created, implementation would be a challenge in India due to a high level of corruption in the permitting process. Like other governments in developing countries in Asia and the Middle and Near East, India's corruption is a sad but true reality.

AHMEDABAD TO AUSTIN

Professor Kurula Verkey in the School of Architecture in Ahmedabad provided valuable guidance and insights for my undergraduate thesis on homes in hot and arid climates. After graduation, I started looking into graduate studies in the U.S. and Europe to continue my research and the professional work I was accomplishing in Kamalbhai's office. I was admitted to the School of Architecture at The University of Texas (UT) at Austin. With one scholarship from a Jain foundation and loans arranged by my father, I obtained financial help to get a U.S. student visa. It was a hard decision for my parents to let me go to the U.S. alone. Their emotional and financial support for future studies was invaluable for me.

I landed at Mueller Airport in Austin, Texas, in the hot month of August 1976. The tiny airport was very close to downtown and the university. Coming from a hot and dry place, Austin's large shade trees were immediately striking, and so began the first of many memorable experiences in Austin—and I had so much to learn about studies, people, culture, food, places, and weather.

THE UNIVERSITY OF TEXAS AT AUSTIN

It was a thrilling experience to join the graduate program in the School of Architecture at The University of Texas at Austin (UT). The university campus has classical stone buildings with red clay tile roofs, outdoor malls, courtyards, and fountains. Tree-covered plazas surrounded the main building, which was the tallest building on campus and known simply as "the Tower." From the south steps of the Tower, there was a direct view of the Texas State Capitol, one of the few downtown buildings of any height visible at that time. The Capitol was by far the most prominent building on the Austin skyline at the time. Downtown Austin today is so full of skyscrapers that the view of the Capitol barely squeezes through some viewing corridors.

The Tower is surrounded by sidewalk promenades, or malls, on all four sides. The South Mall is the grandest of all the walkways, with large oak trees and wide sidewalks on both sides going down to the Littlefield Fountain, another campus landmark that was a gathering place to hang out, do homework, and socialize. The Littlefield Fountain was also a major bus stop for various UT locations. The buses that travel around the university help all students, including new foreign students like me, who were otherwise dependent on walking the campus's forty acres.

SCHOOL OF ARCHITECTURE AT UT

The School of Architecture and the Battle Hall Library face the West Mall and were my two major destinations. Both are stone buildings with clay tile roofs. Battle Hall Library is an iconic building with its winding stairs, tall ornamental ceilings, and quiet spaces for architectural students. Quiet library spaces are a luxury to students coming from India and other developing nations, and this library becomes very special for all foreign students.

The UT campus was grand. By registering for graduate school, one could go to any library on campus and attend any talk open to students. This is the best that American universities have to offer and why students from around the world want to come to the U.S. for higher education. Such an environment encourages open dialogue on all subjects and shows interest in a wide variety of topics. Noteworthy is how students of all ages are enrolled to study. In India, the general belief, especially for women, is that once you complete your college studies, you are done with further education. It was refreshing to me to find older women, moms with children, as my fellow graduate students.

PROFESSOR ARUMI-NOE

Dr. Francisco Arumi-Noe, after becoming a professor in the UT Physics department, decided to move to the School of Architecture. He developed and taught a range of principles and methods, including energy performance and energy analysis of buildings, computer simulation, daylighting, and thermal design. His research had a particular focus on the modeling of passive solar heating and cooling of buildings. Dr. Arumi-Noe was working on these topics long before energy efficiency and building energy performance became issues of public concern.

Under his guidance, as I prepared my graduate thesis, I tested the models of the homes I had studied in Ahmedabad for wind patterns at different times of the day using tunnel experiments at the physics department's laboratory. The university still provides students with the unique opportunity to use facilities like this around campus.

Temperature data that had been collected during my undergraduate thesis was used for experiments. The three-story homes with courtyards provided variations of wind flow and comfort, which determined the behavioral patterns of occupants at different times of day and seasons. I was eager to uncover alternate strategies to keep heat out of buildings, which could be beneficial in homes with or without air-conditioning in hot climates. Plus, for homes with artificial means of cooling, a good design with proper weatherization would save money for occupants, and, more importantly, it would conserve our planet's most important resource, electricity. An added benefit is that when you save electricity, you and others around the globe benefit. Why? Because electricity generation, when done using oil, coal, or natural gas, spews harmful emissions, warming our planet.

PROFESSIONAL WORK IN AUSTIN

I worked in a few prominent architectural firms in Austin like Page Southerland Page and Graeber Simmons & Cowan, both located in downtown Austin. Public transportation was still limited, so like most Austinites, I had to use my car for commuting to work. Austin's population was around 350,000 at the time, and it was easy to travel and park for free around the university and downtown.

I started preparing for the State of Texas's Architectural Registration Exam. In Texas, you can't officially call yourself an architect until you have passed this exam. The exam covered eight subjects over a four-day period and included an eight-hour design test on the fourth day. Besides testing your architectural abilities, it was a test of physical endurance. I managed to pass the exam after one try while I was working at the office of Kinney Kaler Sanders & Crews Architects. I could now call myself a registered architect in Texas and soon became a member of the American Institute of Architects.

Architecture offers many professional paths forward, especially in the U.S. For the first few years after becoming a registered architect, I continued working in private firms on commercial, institutional, and public sector projects. Within a few years, I had an opportunity to work for the City of Austin in the Public Works and Transportation Department, managing building projects. With the city's phenomenal growth came a demand for public projects like libraries, fire stations, health centers, recreation centers, and more. The City of Austin started its major bond programs starting in 1992 after the recession of the late 1980s and the city has continued to grow. The Metropolitan Austin area is spread out, and it has more than 1.2 million people without a well-functioning public transportation system, making a difficult commute for its residents despite the city government's emphasis on sustainable planning, design, and construction.

CHAPTER 5

SUSTAINABILITY AND GREEN BUILDINGS

GREEN BUILDING MOVEMENT OF AUSTIN

The City of Austin owns the 125-year-old electric utility called Austin Energy, which remains a publicly owned utility. The city started promoting energy efficiency and weatherization programs in the 1970s after the nation's energy crisis. Emphasis on sustainable building practices, preservation of the environment, and conservation of energy were driving forces for the establishment of a nationally known ENERGY STAR program in 1992.[1] In 1991, Austin created the first green building program in the nation, rating homes and buildings based on their energy and water efficiency.

This rating system has been developed over thirty-two years, and it is officially called Austin Energy Green Building (AEGB) Certification.[2] Austin's visionary program inspired many cities around the country that were considering ways to incorporate green building standards. It influenced a national organization called the United States Green Building Council (USGBC),[3] which was considering a nationwide green building rating system.

Founded in 1993 by David Gottfried, Mike Italiano, and Rick Fedrizzi, the USGBC created the Leadership in Energy and Environmental Design or LEED rating system to define and measure green buildings. The USGBC issued its first LEED rating system book in 2001. It included a reference guide, credit calculator, and LEED exam study guide. LEED certification provides a framework for healthy, highly efficient, and cost-saving green buildings. It is for all building types, including new construction, renovation, operations and maintenance, and core and shell construction. According to the USGBC, in the 1990s, buildings were responsible for thirty-four percent of the greenhouse gas emissions in the U.S., and that number has now grown to forty percent as the building industry has expanded with growth and urbanization.[4] The USGBC developed a framework and a holistic system to reduce the environmental impacts of buildings, conservation of planet's resources, improve human health, and reduce greenhouse gas emissions to address climate change.

THE FIRST GREENBUILD CONFERENCE IN AUSTIN

It was a watershed moment in the sustainability movement when the USGBC's first international conference and exposition took place in Austin, Texas, in October of 2002. Canadian scientist and environmentalist Dr. David Suzuki was a keynote speaker. His powerful and deeply emotional book *The Sacred Balance* gave concrete suggestions for how we could meet our basic needs and create a way of life that is ecologically sustainable, fulfilling, and just. It offered the seeds of a new direction for civilization, one in which we can all rediscover our place in nature and live in balance with our surroundings. The conference was a resounding success, and I was thrilled to attend it along with my colleagues. The attendance attracted twice as many as expected, and the popular exposition showcased new innovations that included new materials and construction technologies.

AUSTIN'S GROWTH AND PUBLIC FACILITIES

When Austin voters approved building projects in the 1992 bond election, those building projects were in their planning phase. A team of project managers, including myself, in the Public Works and Transportation Department (PWTD) teamed up with Austin Energy to implement sustainable building guidelines in the new bond facilities. The idea was to incorporate sustainable design guidelines in design and specifications to have long-lasting positive impacts on the environment. The guidelines focused on energy efficiency, energy and water conservation, native landscaping, construction waste management, and the selection of low Volatile Organic Compounds (VOC) materials to improve health.

The City of Austin was growing at a fast rate, creating a need for public facilities to serve the citizens. A major bond election was planned in 1998 for voters to approve projects, including museums, libraries, fire stations, recreation centers, building renovations, street reconstruction, and utilities. The project managers in the PWTD would manage these projects on behalf of the city's sponsoring departments like the Library, Parks and Recreation, Public Health, Police, Fire, EMS, Convention Center, Austin Bergstrom International Airport, and others. The USGBC was already working on the LEED rating system. Austin's city management was under pressure to incorporate the LEED standards for the new bond facilities. However, the city leadership was concerned about securing additional resources and training necessary for project managers, owners, and maintenance staff involved in projects.

After discussions, the city council passed a resolution that encouraged 1998 bond projects to incorporate sustainability requirements. Due to uncertainties surrounding the implications of the LEED standards, management did not make LEED standards mandatory. However, they asked the staff to include the LEED rating system, keeping the project within budget.

The LEED rating system v2.1 in the beginning had six major sections: sustainable sites, water efficiency, energy and atmosphere, materials and resources, indoor environmental quality, and innovation.[3] Each area had a point system that allocated credits. Buildings with better performance in each section, especially the energy and atmosphere category, were allocated a higher number of credits. The expected result was an improved performance of mechanical systems, which was a major driver for energy use and, therefore, for climate change.

LEED rating system has developed over a period two different versions and the current version used is LEED v4.1. There are about 200,000 LEED projects in 186 countries and territories. LEED v5 is the newest version of LEED, which is in the review phase. According to the USGBC, LEED v5 is an important milestone in the effort to align the built environment with the Paris Climate Accord's 2030 and 2050 targets. It addresses crucial issues such as equity, health, ecosystems, and resilience.

Project managers who are also architectural or engineering professionals who lead public sector projects funded by cities, counties, states, or federal government have a unique opportunity to represent the owner, especially in defining the project scope and budget. A city project is like having your own project, except it's funded by taxpayer money, adding an extra layer of responsibility. Each public entity has a process to engage the community that it is planning to serve. On a bond-funded senior activity center project in South Austin on Manchaca Road, my project team engaged the senior community in defining the project scope, design, and artwork.

Project managers are fully involved and, at times, are in the best position to provide facts to the management in making decisions to help the citizens of Austin in the long run. Sustainability also means building quality structures while spending public funds wisely.

AUSTIN'S ADOPTION OF LEED RESOLUTION

The director of the PWTD, Sondra Creighton, asked me to join a team that included Matt Curtis in Mayor Will Wynn's office and Richard Morgan of Austin Energy to investigate the implementation of LEED standards for the city's building projects. The PWTD managed most building projects, so it was helpful that our project managers were the owner's representatives, playing an important role in defining project scope that can incorporate LEED building standards. The framework had to be general enough to apply to new buildings, renovations, and maintenance projects.

We were able to draft a resolution that had two major criteria: the cost of construction and the scope of the project as applicable to new construction or renovation projects for achieving a LEED certification. Our team discussed a variety of scenarios for new projects and renovations. We were focused on implementing as many sustainability features in our projects as possible, with or without LEED certification.

In 2007, the city council passed Resolution 20071129-045 (see Appendix 1), requiring LEED certification as applicable in terms of the scope, cost, and type of facility. Our team discussed whether the city facilities should require or have an option to achieve AEGB certification. We decided to get an independent entity like the USGBC to review the city's projects instead of Austin Energy, a city department. This marked a major step forward toward sustainable building practices.

We had many questions after working on the resolution for over a year and determined that all personnel, project managers, city departments, and maintenance staff involved in projects should get training. Soon after the city council passed the LEED resolution in November 2007, PWTD took the lead and formed an interdepartmental sustainability working group of relevant department representatives to help with the implementation of this resolution. The city management created the Office of Sustainability[5] a few years after this resolution passed. It oversees reviewing operations of all city departments to reduce environmental impacts and lower emissions by focusing on equity and justice issues for communities around the city.

The new International Energy Conservation Code (IECC) of 2021, which was adopted by the City of Austin, requires the commissioning of all cooling and heating systems and sealing air and water leaks to conserve energy. A new code's adoption requires the dissemination of information to design teams, so they know its implications. Austin Energy has been organizing helpful educational training seminars for Austin's design and construction community for a long time. It also holds green builder training classes. These programs by Austin Energy make it a unique utility serving our community.

Building codes are the best way to reduce carbon emissions resulting from building activity while addressing the climate crisis. The vernacular architecture discussed in Chapter 2 provides a historical perspective on finding comfort in buildings. The incorporation of passive climate elements not only creates pleasant environments but also reduces the need for heating and cooling, thereby reducing harmful emissions. In April 2024, President Biden rolled out new minimum energy standards for affordable housing using federal funds. The new 2021 International Energy Conservation Code standard, which applies to single-family homes and small apartment buildings, is thirty-four percent more efficient than the 2009 edition used by the Department of Housing and Urban Development (HUD) and the Department of Agriculture (USDA).[6] Energy-efficient buildings are good for the users and good for the planet.

ADVANCING SUSTAINABILITY IN CITY PROJECTS – BUILDINGS AND INFRASTRUCTURE

LEED FACILITIES

As of February 2024, the City of Austin has forty-six LEED-certified buildings, as shown below:

- 2 LEED Platinum—Austin Central Library

- 22 LEED Gold

- 18 LEED Silver

- 4 LEED Certified projects

Currently, there are forty-five city projects in progress that are pursuing a rating system. Austin's resolution requires minimum LEED Silver certification if the project scope and budget requirements stated in the resolution are met. Often, teams were able to achieve Gold certification when they were trying for Silver.

According to Austin Energy Green Building, Austin has fifty-nine rated commercial buildings including residential high-rises—forty-two of them also have AEGB star-rating. For multi-family buildings, that are seven stories or less, there are fifteen projects that have AEGB star ratings. Educational institutions like The University of Texas, Austin Community College, school districts area-wide, and privately owned buildings have LEED and/or AEGB ratings that are not counted here.

THE PFLUGER PEDESTRIAN BRIDGE

The pedestrian and bicycle bridge project called the Pfluger Bridge in Austin was completed in 2001. The project is not certified by any rating system, but it was an infrastructure project that was able to enhance the surrounding environment and provided a safe and beautiful facility that helped improve the health and well-being of Austinites.

This project was originally funded in the 1984 bond election to widen the Lamar Bridge and was being managed by the transportation division of PWTD. They did feasibility studies on widening the narrow sidewalks on both sides by adding overhangs. The historic Lamar Bridge formed a west boundary for the city and opened to Army tanks and trucks in 1942, when our city's population was close to 90,000. By the early 1980s, the Lamar Bridge was a major artery connecting the north and south sides of the city. Increased vehicular and pedestrian traffic prompted PWTD to look for options to expand the unsafe three-foot-wide sidewalks. If you had ever walked on these narrow sidewalks, you would remember how nerve-racking they were as you faced walkers and runners with their dogs trying to cross the bridge. For users of the hike-and-bike trails, this was the only crossing

Fig. 1. The three-foot-wide sidewalks on the Lamar Bridge that crossed Town Lake were unsafe but heavily used. PWTD started a feasibility study to expand the bridge's sidewalks after a fatal accident involving a pedestrian.

near Lamar Boulevard.

Lady Bird Johnson, an avid environmentalist, had promoted the hike-and-bike trail system around Town Lake, later renamed Lady Bird Lake, which remains one of the most attractive features of Austin. Without enough north–south bridge crossings over Town Lake, the PWTD was under pressure from users of this bridge to widen the sidewalks.

In engineering studies, the city had looked at overhanging the sidewalks from the historic bridge. They had sketches done, showing large overhangs to the bridge on each side. But this graceful open-spandrel concrete-arched bridge was under the jurisdiction of the State of Texas. Any modifications to the historical structure would require approval of the Texas Historical Commission. The Texas Historical Commission was concerned about a cantilevered addition to an old structure along with its possible structural and visual impacts. They didn't approve the modifications.

The PWTD's transportation division worked on this project for more than ten years and found no pathway to move forward. Then, Director Peter Rieck asked me to take over the project. The only solution was to construct an independent structure. However, an independent structure on Town Lake was unthinkable for some prominent Austin architects and urban designers who remained opponents of this idea. A popular rowers' club was concerned about the pathways in the otherwise open vistas of the lake. Our team studied all the concerns of different users, keeping in mind three major design goals: safety, environment, and aesthetics. The bridge structure we designed had a safe crossing over the lake, clear pathways for rowers, and aesthetically pleasing support bents, upper structures, railings, lighting, and surfaces to enhance the beauty of Town Lake and protect the beautiful trees. We then reached out to the Austin community, including artists, and received a lot of input on the design.

During the design phase, a pedestrian on the sidewalk of the Lamar Bridge was killed by a motorist. PWTD management wanted to move forward as quickly as possible with bidding and construction of a new bridge. We had to bid on this project twice due to higher bids. Our progress was getting bogged down. In a meeting with the city manager regarding the two bids, I made the case to build a main bridge that would address the closing of the narrow, unsafe sidewalks of the old Lamar Bridge since safety was of paramount importance. Removing an extension over Cesar Chavez from the scope of work would bring the main project within budget. With that suggestion, City Manager Jesus

Fig. 2. In 2001, the new Pedestrian and Bicycle Bridge over Austin's Town Lake transformed the surroundings. This photo is from the south side of the lake, showing the beautiful structure of the historic Lamar Bridge. The weathering steel material used for large, curved beams holding the upper structure of the new bridge was made from recycled steel, requiring hardly any maintenance. The concrete bents were carefully designed to maintain the aesthetics of the jewel that is Town Lake. Photo Credit: HDR/Paul Cockrell. HDR Engineering was a prime consultant for this project with a team of sub-consultants.[7]

Fig. 3. Bird's-eye view showing the helix at the north end, a prominent feature of this bridge. The entire bridge is accessible to people in wheelchairs and bikers. The extension, which was initially removed from the scope, is now completed.
Photo Credit: HDR Engineering/White Photography

Garza and Assistant City Manager Toby Futrell decided to move forward and seek the City Council's approval. The professionals who were opposed attended that meeting of the City Council, which lasted past midnight. They spoke in opposition, and the Texas Historical Commission's representative spoke in favor. In the end, the City Council unanimously approved the construction contract at 2:00 AM one early April morning in 2000.

The hike-and-bike trails and the streets on both sides were kept open during the entire construction duration. The whole design team and the contractor partnered well to focus on the completion of this bridge in one year despite extremely wet weather during construction.

The joy was apparent on the grand opening day in 2001 when thousands of people showed up to celebrate with dignitaries, including Austin Mayor Kirk Watson and U.S. Congressman Lloyd Doggett. The bridge has become a great park-like place

Fig. 4. In addition to being a pedestrian and bicycle bridge, it provides a park-like atmosphere with excellent views.
Source: Photos from the author's collection

to walk, bike, relax, watch the sunset, and enjoy the surroundings. Besides walking and running, I have seen people getting married, having book readings, doing yoga, playing with pets, and teaching biking to young kids on the bridge, which still provides a north–south connection to bikers. The Austin community has fully embraced the bridge, and it warms my heart whenever I hear from friends about its special place in their lives.

The extension of the bridge that connects the north-side developments on Cesar Chavez Boulevard was completed in 2010. Pedestrians and bicyclists have seamless access, providing and encouraging alternative transportation from the south to the north.

Community engagement is crucial in city projects. As a project team with sustainability as one of our major goals, we had to manage competing demands on scope and budget, and schedule and make the opening day of the Pfluger Pedestrian Bridge a memorable day. We demonstrated that when all parties understand a project's goals, we, as green builders, can get the best possible outcome despite the given constraints.

GEORGE WASHINGTON CARVER MUSEUM
AND CARVER LIBRARY

The George Washington Carver Museum and the Austin Public Library's (APL) George Washington Carver Branch in East Austin at Angelina and Rosewood Streets were two 1998 bond projects approved by voters. APL's Carver Branch had long been a focal point for the residents of East Austin. The Black community had a small museum in a tiny historic house at this site. For the first time in the city's history, a major museum project was funded to showcase the East Austin community's contributions.

A robust public engagement process was conducted during the feasibility study phase. The interstate highway, I-35, divided East Austin from downtown, not only physically but also in terms of economic development. The difference was stark when you crossed I-35 to the east prior to 1998. The gentrification of East Austin has driven out many families who had lived there for a long time.

Fig. 5. The George Washington Carver Museum at 1167 Angelina Street. Carter Design Associates was a prime consultant on this project with a team of sub-consultants.[8] Photo: Author's collection

A LEED rating system was not mandatory for these projects. APL's Carver Branch expansion added services for a children's reading room and computer tutoring lab, a priority for the community. The Carver Museum's scope included Dr. George Washington Carver's science lab for children, a dance studio, an archival space, and an auditorium. The museum's permanent exhibit was designed to tell the story of Juneteenth, a celebration to commemorate the day the last enslaved Black Americans learned that they had been emancipated. Since it originated in Galveston, Texas, on June 19, 1865, Juneteenth has been a special day of celebration for Texans.

In 2005, this beautiful facility opened its doors to the public with a joyful day of celebration of the city's first Black cultural facility and an expanded Carver Branch library. Although not mandatory, we achieved a LEED certification.

The Carver Museum and Carver Branch renovations were my team's first two LEED-certified projects under version 2.0. The experience was educational for us. The city hired an independent commissioning agent's service for testing all systems, specifically heating, ventilation, and air-conditioning systems (HVAC) for intended performance. The biggest lesson we learned was the importance of tightly sealed building walls, windows, roofs, and connections. If the building enclosure is not sealed, tested, and commissioned, moisture in Austin's humid climate becomes a major issue. Dehumidifiers were part of the mechanical system for controlling humidity, but they increased energy consumption, making it difficult to gain energy and atmosphere credits under LEED v2's rating system.

The city of Austin has now adopted the IECC 2021 and UBC 2021, which require the sealing of a building's exterior skin elements to keep moisture out. This reduces the use of an HVAC system and saves energy. With global warming, the design, installation, and performance of an efficient HVAC system have become critical building components.

AUSTIN ANIMAL CENTER

Fig. 6. Austin Animal Center, 7201 Levander Loop, was completed in 2011. I managed this project, which achieved a LEED Gold Certification. The design firm of Jackson and Ryan Architects was a prime consultant with a team of sub-consultants.[9]
Photo Credit: Mark Scheyer and Jackson & Ryan Architects

Austin Animal Center is located on a campus that was home to the Texas Blind, Deaf, and Orphan School for people of color from 1889 to 1961. It was important to maintain the quadrangle in front of the building, with sidewalks around it, to conserve the campus's history. We reused bricks from demolished buildings, as shown in the picture above. On this very flat site, we used pervious concrete pavement for sidewalks, pervious pavers for parking lots and plazas, and rain gardens to alleviate stormwater drainage issues. Other energy-saving features include natural light, solar panels, white roofs, energy-efficient lighting and controls, an efficient mechanical system design for energy savings of thirty-two percent, water-conserving fixtures, and a special cleaning system to reduce water consumption by thirty-five percent. This site is full of heritage trees, which were preserved for visitors to enjoy while visiting the Animal Center.

AUSTIN CENTRAL LIBRARY

Austin Public Library's Central Library, another LEED project, is an outstanding public facility located across from Lady Bird

Fig. 7. New Central Library building of Austin with LEED Platinum certification. Lake Flato Architects were the prime consultants with a host of sub-consultants on this project.[10]
Photo credit: Cynthia D. Jordan, Project Manager and colleague of the author

Lake. It provides beautiful views of Town Lake, now renamed Lady Bird Lake, and the city. The six-story building surrounds an atrium with stairs in the middle, allowing natural light to penetrate the interior, creating interactive interior spaces. The project team made use of many sustainability features, including a variety of energy-reduction strategies, solar panels, indoor and outdoor water reduction, and rainwater collection. Most importantly, the library has become a destination for Austinites and visitors, making it a fun learning place for people of all ages. This library project was the first City of Austin's LEED Platinum-certified facility, the highest certification by LEED.

AUSTIN ENERGY'S NEW HEADQUARTERS

The Mueller Redevelopment is a unique pedestrian-friendly place where it is possible to live, work, buy food, and play within the developed area. Such developments reduce need for driving and reduce emissions. The land used to be Austin's Mueller Airport, before the city's airport moved to the former Bergstrom Air Force Base. Austin Energy's new headquarters building is in the Mueller Redevelopment. It is an exemplary structure with three certifications: Austin Energy Green Building-Level 5, LEED Platinum, and WELL Gold Certified.

Among Mueller's many sustainability features, a few stand out:

- As part of WELL's rating requirements, this building provides fruits and vegetables and information on ingredients. Eating areas for employees are located such that they enhance meetings among coworkers since healthy diets have the potential to nurture human health and reduce overall polluting emissions.

- This building places an emphasis on showcasing local artists and environmentalists.

- This building has natural daylight for most locations with automatically controlled shades.

- This building implemented a whole-building life cycle assessment to make informed design decisions that ultimately reduced the global warming potential of building materials by more than eleven percent.

- This building used an energy model for the construction of the building, which helped earn sixteen points for energy efficiency in the AEGB rating. This means that this building is performing sixty-five percent better than a code-built building.

OTHER SUSTAINABILITY CERTIFICATIONS

Austin Energy Green Building program, a leader in the 1990s, continues to work with Austin's design and construction community. Its mission is "to cultivate innovation in building and transportation for the enrichment of the community's environmental, economic, and human well-being." The AEGB's rating system rewards sustainable building practices, leads to high-performing buildings, and creates market demand for green buildings. As the market recognizes sustainable building practices, these measures are incorporated into Austin's codes and local building regulations. AEGB still provides training programs and seminars on a regular basis to carry out its mission. AEGB is a tremendous asset for the Austin community, and that work is possible because of the city's commitment to energy-efficient, sustainable buildings.

The Living Building Challenge focuses on maximizing the positive impacts specific to the location, community, and culture of a project. According to the International Living Future Institute, living buildings are "regenerative buildings that connect occupant to light, air, food, nature, and community," "are self-sufficient and remain within the resource limits of their site," and "create a positive impact on the human and natural systems that interact with them."

I applaud this philosophy and framework because it can be transformational. The Institute's certification program was recently started, and version 4.0 was released in June of 2019.[11] "It is a tall order," as they admit, and difficult for most owners to achieve. But it is a gold standard to strive for wherever it is doable.

Josey Pavilion,[12] built by Lake Flato Architects in Decatur, Texas, is the first project to be certified by the Living Building Challenge in the State of Texas. This pavilion is self-sufficient, providing its own water and energy with the least impact on the natural realm. The City of Austin has not yet pursued this challenge in its buildings, but it continues to strengthen its building code requirements.

The International WELL Building Institute is a global movement. The WELL Building Standard® is a performance-based system for measuring, certifying, and monitoring features of a built environment that impacts human health and well-being through air, water, nourishment, light, fitness, comfort, and mind. It has been developed over the course of ten years and is backed by the latest scientific research.[13] The City of Austin's Planning and Development building located at 6310 Wilhelmina Delco Drive received its WELL Gold certification in 2022.

EVOLUTION OF SUSTAINABILITY IN AUSTIN'S PUBLIC PROJECTS

The city has an interdepartmental team called Built Environment Equity and Sustainability, or BEES, that is a cross-departmental/partner organization hub for collaboration and coalition. Prior to being named BEES, it was called the Interdepartmental Sustainability Working Group (ISWG) when I provided leadership in the formation of the group at its inception in 2008. BEES aims to break down barriers and provide effective channels to advance equitable, sustainability-focused, place-based assets and operations of the City of Austin. The LEED resolution of 2007 was revised by combining all previous resolutions, and the latest resolution (20210902-042) was a comprehensive document to provide guidance for a green building policy for all types of projects using different methodologies for planning, designing, and constructing new buildings and performing renovations (see Appendix 2).

The idea is to have all projects—new buildings, major renovations, small projects, civil and infrastructure projects—incorporate sustainable design features and construction specifications in structures built on city-owned land. The Office of the

City Engineer has been implementing sustainability in all its in-house street and infrastructure projects and has found innovative ways for pavement and surfacing that would have positive impacts on the environment.

BEES encourages all or any rating system that is applicable to projects managed by the city's project managers:

1. BREEAM Infrastructure—formerly known as CEEQUAL—is the international evidence-based sustainability assessment, rating, and award scheme for civil engineering, infrastructure, and landscaping in public spaces. (ceequal online.com; bregroup.com)

2. The Institute for Sustainable Infrastructure offers Envision, a set of guidelines that aid in optimizing the sustainability of an infrastructure project during the planning and preliminary design phases, as well as the means to quantify the relative sustainability of the project. (sustainableinfrastructure.org/envision/use-envision)

3. The Green Road Rating System is basically an instruction manual for measuring and managing sustainability on infrastructure projects like streets, highways, bridges, rails, and trails. It provides critical guidance for cities, counties, states, and countries seeking a resilient and sustainable future. (transportcouncil.org)

4. Parksmart offers a lifetime of returns on parking structures through reduced operational costs, increased energy efficiency, and better lighting and ventilation. It is the world's only certification system designed to advance sustainable mobility through smarter design and operation of parking structures. The City of Austin's parking garage at Austin-Bergstrom

International Airport has achieved the Parksmart certification. (parksmart.org)

5. SITES provides a comprehensive framework for designing, developing, and managing sustainable and resilient landscapes and other outdoor spaces. SITES is the ideal tool to support a nature-positive design. (sustainablesites.org/)

Zero-waste initiatives in building design are already being incorporated in some buildings. Waste generation is responsible for five percent, as of 2016, of all emissions that are warming our planet.[14] A new certification called TRUE (Total Resource Use and Efficiency; rubicon.com/blog/true-certification) was recently introduced and incorporates aspects of waste reduction, not only after a building is occupied but also in design and the selection of materials.

For all professionals in the building industry, it is important to understand a whole building's life-cycle assessment, from preliminary design, material selection, final design, construction, operation, and maintenance to the end of a building's life. Operations and maintenance account for the largest amount of greenhouse gas emissions in a building's life-cycle process for air-conditioned and heated buildings. We professionals must employ all available strategies, including passive design elements and mechanical designs, to reduce the carbon footprint of buildings.

CHAPTER 6

BUILDINGS' IMPACT ON CLIMATE CHANGE

BUILDINGS ARE SPECIAL

Human beings have a special emotional connection to buildings, whether homes, places of work and worship, schools and colleges, shops and malls, libraries, museums and theaters, health and recreational facilities, or industrial complexes. The list is long. We want comfort and joy in those places, and that's been true for centuries.

Even before electricity, civilizations around the world built remarkable buildings. Once fossil fuels were found and electricity was discovered, the growth of buildings along with modern conveniences for heating and cooling has been phenomenal. With technological innovations, population growth, and high-consumption lifestyles, the world's electricity consumption has skyrocketed along with the emissions resulting from electricity generation, transportation, manufacturing, agriculture, and all our activities.

WHAT ARE GREENHOUSE GASES?

So, how does the burning of fossil fuels and spewing emissions affect our environment? Gases that trap heat in our atmosphere are called greenhouse gases (GHGs). Think of invisible clouds in

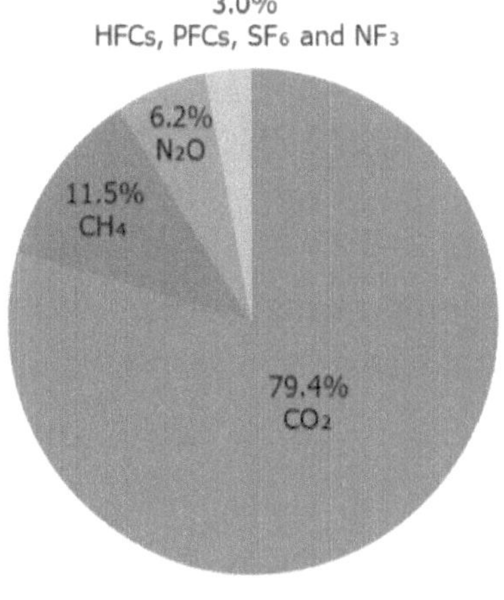

Fig. 1. U.S. Emissions in 2021. (epa.gov/ghgemissions/overview-greenhouse-gases) Carbon Dioxide is CO_2, CH_4 is Methane, N_2O is Nitrous Oxide.

U.S. Environmental Protection Agency (2023). Inventory of U.S. Greenhouse Gas Emissions and Sinks: 1990–2021

the air. They are made of carbon dioxide, methane, nitrous oxide, and fluorinated gases. Carbon dioxide consists of 79.4 percent of all GHGs, according to the EPA.[1] See above.

Carbon dioxide (CO_2) is a colorless and odorless gas that stays in our atmosphere for hundreds of years, and that is why it is concerning. Currently, according to the Scripps Institute of Oceanography, more than 1,200 gigatons of carbon have accumulated in the atmosphere[2]. Just to give an idea—one gigaton is equal to about 2,204 billion pounds! Though not visible to the naked eye, CO_2, along with other GHGs, traps heat, increasing global warming. According to the Global Monitoring Laboratory of NOAA, "The amount of carbon we are putting into the atmosphere each year is equal to twenty times the weight of the current world population."[3] We continue to add carbon dioxide each year.

So, what do we need to do? We must drastically reduce the carbon footprint in all sectors and find ways to remove excess carbon dioxide that has already accumulated!

Climate scientists around the world are hard at work on building models and doing analyses to show us that this excess carbon dioxide is causing the average global temperature to rise, changing Earth's energy balance, which results in catastrophic extreme weather events. As I shared in Chapter 1, the number and intensity of such events have increased substantially in the five years from 2016 to 2021, as compared to the forty-one years between 1980 and 2021.

HEAT STRESS

The effects of heat on the human body have been extensively studied by health professionals and the U.S. Center for Disease Control. Baruch Givoni, in his 1969 book *Man, Climate and Architecture*, gives a single formula called the Index of Thermal Stress (ITS)[4] that describes the mechanism of heat exchange between the human body and the environment and from which the total thermal stress on the body, metabolic and environmental, can be computed. Besides considering temperature, radiation, humidity, and wind velocity, this formula considers the type of clothing and the terrain.

During events of extreme thermal stress in the absence of any cooling, the human body is unable to cope. Heat-related illnesses, even deaths, occur in some cases. Heat-related deaths have been on the rise, and that has prompted many cities, including Miami, Florida, Los Angeles, California, Phoenix, Arizona, and Athens, Greece, to appoint a heat officer. The city of Phoenix had thirty-one consecutive days above 110 degrees Fahrenheit in the summer of 2023. Think of the effects of such heat on outdoor workers! The Biden Administration has created a website called heat.gov that provides the most current conditions and some coping mechanisms.

BUILT ENVIRONMENTS

Professionals who design and construct buildings and infrastructure projects have a whole new set of environmental conditions to consider in their projects. According to architecture2030.org, built environments generate forty-two percent of annual global CO_2 emissions. Of total emissions related to built environments, building operations (that includes heating, cooling, lighting, and other uses) are responsible for twenty-seven percent annually, while building infrastructure materials and construction, typically referred to as embodied carbon, are responsible for an additional fifteen percent annually. Buildings are dear to us, and we spend a lot of time in them. Therefore, professionals in this industry have a responsibility to find ways to design and construct as sustainably as possible for both new and existing buildings.

By making decisions regarding green buildings in the early planning phase, planners, designers, architects, and project teams have many opportunities to implement all possible passive climate strategies, as well as modern innovative strategies to lower the carbon footprint throughout the life cycle of a project.

Once completed, buildings last for many years, creating a huge stock of existing buildings. Billions of people who use existing buildings continue to use largely fossil fuel energy for cooking, lighting, heating, and cooling. Renovating, repurposing, and improving the energy efficiency of existing buildings instead of demolishing them would help tremendously to reduce carbon emissions.

Besides the repurposing and constructing of green buildings, it is important for these same groups of professionals to address our overwhelming use of fossil fuels for transportation, energy generation, industrial production, and agriculture. Renewable energy's growth is phenomenal, but we are still largely dependent on fossil fuels.

EMISSIONS BY SECTORS IN THE U.S.

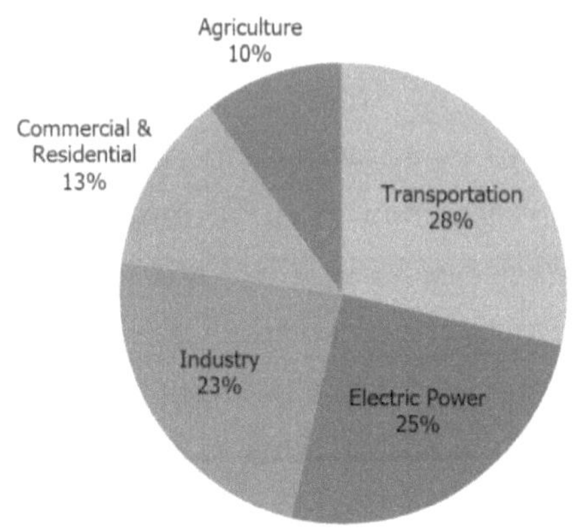

Fig. 2. Total U.S. greenhouse gas emissions by economic sector in 2021. Source: The U.S. Environmental Protection Agency (EPA)

- Twenty-eight percent of emissions come from transportation, mostly from burning petroleum (gasoline and diesel) in cars, trucks, ships, trains, and planes.

- Twenty-five percent of emissions come from electricity power generation; sixty percent of our electricity comes from burning coal and natural gas.

- Twenty-three percent of emissions come from burning fossil fuels for energy and the production of raw materials.

- Thirteen percent of emissions come from businesses and homes, as the result of burning fossil fuels to heat and cool buildings.

- Ten percent of emissions come from livestock and crop production.

Land use, changes to land use, and forestry in the United States amounted to a net reduction. They offset twelve percent of greenhouse gas emissions in 2021, according to the EPA. Protecting forests, planting urban trees, creating gardens, and using all possible land surfaces wisely and equitably can do a lot to reduce future emissions.

During the planning and design phases, project teams must make important decisions about orientation, site conditions, floodplains, local materials, and the energy-efficient envelope. These teams should select the best-performing HVAC systems during the preliminary planning phase.

In the globalized world, building materials can be extracted in one area, manufactured in another, stored in yet another location, and then transported to construction sites. Embodied carbon of building materials is defined as the carbon emissions released from extraction, manufacturing, transportation, and construction. We are just beginning to count the embodied carbon in the life cycle of materials. To reduce the embodied carbon of materials, market transformation needs to take place with effective policies. Buildings can also generate on-site electricity, reducing electrical generation demand on the power grid. To stabilize and cool our climate, we not only need to reduce embodied carbon from buildings and other sectors but also remove carbon dioxide that has accumulated in the atmosphere for many years.

LIGHTING IN BUILDINGS

Research and innovations to make light bulbs that can stay lit had been ongoing in Europe and the United States in the early nineteenth century, showing human ingenuity. Innovators continued their work, but the trick was to have light bulbs that could stay lit for an extended period. In 1879, Thomas Edison invented the first such continuously lit incandescent bulb, which was critical for use in buildings. In 1882, the Edison Electric Illuminating Company of New York brought electric lights to parts of Manhattan.[5] Progress was slow in building power plants and transmission lines. For the next fifty years, most Americans still lit their homes with gas lights and candles. Thereafter, electricity became an important feature in households in urban areas around the world.

Innovations of light bulbs focused on low energy consumption and spreading a more pleasing light continued for years, and research is still ongoing. From improved incandescent bulbs to fluorescent bulbs to energy-efficient fluorescent bulbs to compact fluorescent light bulbs of all varieties and, finally, to the latest, light-emitting diodes—better known as LEDs—which are taking over the light fixture industry and have gradually reduced electricity consumption while improving quality of light over time. People in developing countries, where the 24/7 availability of electricity remains a dream, understand the value of this precious commodity.

Lighting impacts lifestyles in a major way. Children who use streetlights to study are surprisingly common in certain villages in India, Africa, and other parts of the world. Working families benefit from light bulbs by having additional time to work at night. Buildings designed with LEED and other sustainability rating systems have emphasized the use of natural light and a reduction in electricity usage to reduce emissions from fossil fuels. Electricity is measured in kilowatt-hours (kWh). Assuming fifteen cents a kilowatt-hour of energy, the savings when comparing three types of light bulbs are considerably different, according to arcadia.com:

Incandescent bulbs: A 100-watt bulb running for a full year would use 876 kWh of energy, would cost $131.40, and require twelve replacements over one year.

Compact fluorescent light (CFL) bulbs: To match the brightness of a 100-watt incandescent bulb, you would need a 25-watt CFL bulb. It would use 216 kWh of energy, would cost $32.40, and would require two replacements per year.

<u>LED (Light Emitting Diodes)</u>: For the same brightness, you would need a 16-watt LED bulb. It would use 140 kWh of energy, would cost $21, and would not require any replacement over one year.

Lowering electricity use by light bulbs means lowering the cooling load in buildings, saving precious electricity. Check out your electric bill and see how many kWh of energy your family uses and find ways to reduce your electricity usage and bill.

COOLING DEVICES IN BUILDINGS

The first noted ceiling fan was found in India in the seventeenth century, and its Indian name is *punkha*. Punkha were made with palm leaves or cloth and hung from the ceiling. Handheld fans were used by ancient cultures and are still used in many parts of the world. Around 1860, the first ceiling fan that did not use electricity was installed in the United States. It was run using a belt system, water, or steam energy. The first electric ceiling fan in the United States was invented in 1889 by Philip Diehl,[6] who worked in a sewing machine company and used a sewing machine motor. In large parts of the world, ceiling fans are major equipment for comfort from high temperatures when air-conditioners are not affordable or when locations do not have enough electricity to run them. I know I craved fans in my childhood during summers. Why are ceiling fans still critically important? When used in concert with the air-conditioners, they provide comfort at a slightly higher temperature, saving electricity and cutting down emissions.

In the 1840s, physician John Gorrie of Florida proposed an idea of cooling cities to relieve residents of "the evils of high temperatures."[7] He believed that cooling was the key to avoiding diseases like malaria and making patients more comfortable.

Gorrie's work laid the foundation for modern air-conditioning and refrigeration.

In Europe, Carl Linde started working on his ice machine company in the early 1870s. In 1894, he founded a company that developed one of the greatest products in human culture—summer beer. He was knighted as Ritter von Linde for his lifelong achievements in Bavarian brewing technology and refrigeration in 1897.

Years after Dr. John Gorrie and Dr. Carl Linde's work, air-conditioning was first invented in 1902 by a skilled New York engineer named Willis Carrier, who experimented with the laws of humidity control in his printing plants. Carrier designed a system using cooling coils and received a patent for what he called an "Apparatus for Treating Air,"[8] which could either humidify or dehumidify air. At the St. Louis World Fair in 1904, organizers used mechanical refrigeration to cool the Missouri State Building, including a 1,000-seat auditorium.[9] It marked the first time that the American public was exposed to the comfort of cooling by mechanical equipment.

Theaters were the first public places to provide cool shelters for entertainment. In the late 1960s, central air-conditioning combined with heating to create what is commonly known as an HVAC (heating, ventilation, and air-conditioning) system. There is a fascinating history of air-conditioners documented by ASHRAE.[10] Back then, U.S. homes and small facilities used small window air-conditioning units for convenience, cost, and conservation of electricity. With an increase in population, urbanization, economic growth, and the development of energy sources, our combined energy consumption is rising at a fast pace. Homes are becoming larger and larger in the U.S. and India. As a result, the need for HVAC equipment is also rising, as is our use of electricity to run those systems.

My nephew's children, who go to an elementary school in

Metuchen, New Jersey, had to leave their school by midday from September 5–8, 2023, because it was unusually hot in their classrooms without air-conditioning. My relative in New Delhi, who had taught in a high school for many years in classrooms without air-conditioning, decided to retire, and among other reasons, the health effects of heat played an important role. In Ahmedabad, the rise in temperatures, along with an increase in humidity, has made the heat harder to tolerate without air-conditioning. Those who can afford the expense are ramping up their use of cooling devices.

REFRIGERANTS IN AIR-CONDITIONERS

Refrigerants play a key role in air-conditioners to cool the circulating air for the comfort of inhabitants. Chlorofluorocarbons (CFCs) and hydrochlorofluorocarbons (HCFCs) were used as refrigerants in A/C systems for a long time. Scientific research showed that CFCs and HCFCs caused a depletion of the beneficial ozone layer.[11] The sky's ozone layer, also known as stratospheric ozone, is good ozone that is essential for absorbing the sun's ultraviolet radiation, which is harmful to human health. In 1987, nations of the world produced an international agreement called the Montreal Protocol,[12] which was an important and effective agreement—a gold standard when it comes to global agreements—to phase out harmful refrigerants like CFCs and HCFCs to help the sky preserve Earth's helpful ozone layer that protects our health.

Their replacement refrigerant was a chemical called hydrofluorocarbons (HFCs). This refrigerant was not as harmful as CFCs and HCFCs, but its capacity to warm the atmosphere was 1,000-9,000 times greater than that of carbon dioxide. Global warming was already recognized as an important issue. In

October 2016, an agreement called the Kigali Amendment[13] to the Montreal Protocol was signed by most nations to help phase out HFC, a very harmful refrigerant.

The U.S. didn't sign it because President Trump was not ready to sign the Kigali Amendment. Members of Congress were under pressure from U.S. manufacturing companies, which were already planning to phase out HFCs to compete with manufacturers outside the U.S. The manufacturing companies were aware of the move by other nations to phase out HFCs per the Kigali Amendment and were concerned about their market share in selling those products. The U.S. Congress, with support from members of both parties, eventually signed the Kigali Amendment as part of a budgetary agreement. This is an example of public actions and market transformation. The richest countries of the world have started phasing out this planet-warming refrigerant, while developing countries have until 2024–2028 to phase out HFCs per the Kigali agreement. In developing countries like India and China, the use of A/C has gone up substantially and so has the use of HFC refrigerants, adding to global warming.

In the northwestern part of the U.S., installing A/C in homes was not a common practice. A higher latitude, cool and humid weather, large pine trees, and heavily wooded areas helped to maintain their low temperatures, even during summers. Wildfires have destroyed some of these wooded areas, and the overall rise in global warming has affected summer temperatures there. When this region started experiencing much higher temperatures in 2011, homeowners started installing air-conditioners if they could afford them. My stepson has a house with large glass walls. It was not designed for A/C back in the day, but he installed A/C units that year. Those who did so felt lucky during July 2021, when an unprecedented heat wave hit this area, causing deaths and destruction.

AUSTIN'S DOWNTOWN AND URBAN HEAT ISLAND EFFECTS

Downtown Austin is now full of glass skyscrapers, causing what is being called an urban heat island effect, which is created by tightly packed buildings, a lack of green spaces, and the use of air-conditioning for millions of square feet of space in a concentrated area. Temperatures in the downtown area could be five to seven degrees higher than otherwise due to this urban heat island effect, according to Climate Central.[14]

Austin's expanded summer season and record-breaking high and low temperatures in summers are affecting people with heat-related health issues and causing a rise in deaths in some cases. According to David Yeomans, former chief meteorologist of KXAN News in Austin, "As the climate changes, Austin's average high during the hottest part of summer has gone from ninety-seven degrees to ninety-nine degrees (Fahrenheit) just in the twelve years that I've worked here. Soon, 100 degrees (Fahrenheit) will be the average for the hottest part of summer."

Fig. 3 - Left. A view of tall glass buildings of Austin's downtown from the Central Library. Source: Photo from the author's collection

Fig. 3 - Right. A view from south of Lady Bird Lake, Photo credit: Toni Thomasson

The use of air-conditioning goes up as it gets hotter. It is a vicious cycle: increased heat requires more air-conditioning, which uses more electricity and refrigerants that increase emissions that worsen global warming for our planet.
With that, we jack up the air-conditioning.
And the cycle continues! This is not sustainable.
The only solution, then, is to address global warming, reduce emissions, and restore our climate.

SINGAPORE'S URBAN HEAT ISLAND EFFECTS

During my visit to Singapore in the summer of 2022, I found something unique that the government was doing to cool the environment. This island city-state is full of high-rise towers that require enormous amounts of cooling due to the location's warm and humid climate. Land is scarce. But, according to naturalwalkingcities.com, 46.5 percent of Singapore's land in 2020 was covered in green space, with a tree-canopy percentage of almost thirty percent.

Austin residents have made a difference in air quality by using hybrid cars, electric vehicles, bicycles for commuting, single-stream recycling, and composting. We could do better with thoughtful policies for public transportation to reduce tailpipe emissions. According to the latest census numbers, Austin is the eleventh largest city in the U.S., but it lacks an effective mass transit system for public transportation. Its traffic problems are likely to persist, worsening the air quality. Anyone who drives on Interstate Highway 35 (I-35) witnesses this condition day after day. In addition to maintaining its tree-canopy coverage, Singapore focuses on public transportation to reduce concrete streets and private motor vehicles.

For cities in the U.S., New York has the most amount of green space, which is twenty-five percent. Austin places second with fifteen percent, according to treehugger.com.[15] Fortunately,

Fig. 4. The majority of Singapore's closely packed high-rise towers are fully air-conditioned, with mechanical units on the roofs or sides of buildings. Green spaces are created using every inch of available building surface or piece of land.
Sources: Photos from the author's collection

our early city leaders had the foresight to purchase land for parks throughout the city. The biggest single park in downtown Austin, 358-acre Zilker Park Complex, which includes Barton Springs, was given to the city's inhabitants by Andrew Zilker in trust to be kept and maintained for the city's people. In Ahmedabad, new developments for apartments and office space are being constructed by removing mature trees, worsening the urban heat island effect.

Singapore is a city-state that is located on a small island. Sustainable building construction and the extensive use of gardens supported by governmental policies have made a huge difference in absorbing carbon dioxide to reduce emissions. Still, there are pockets of the downtown that can have temperatures five to ten degrees Fahrenheit higher than the surrounding areas.[16] To combat the heat, the government is introducing additional measures like painting the roofs with light reflective coatings, passive climate designs for individual buildings, and improving the layout of streets for wind circulation to cool the streets and reduce the urban heat island effect. Singapore's wealth and centralized political system have allowed it to make decisions that can be implemented quickly. It has a well-planned infrastructure and uses sustainable design elements, water conservation, planting on vertical and horizontal surfaces of building facades, roof gardens and terraces, and efficient air-conditioning systems to reduce the use of electricity to cut down emissions.

GLOBAL RISE IN A/C USE AND GLOBAL EMISSIONS

The latest results from the Residential Energy Consumption Survey (RECS) show that at least ninety percent of U.S. households used A/C in 2020.[17] The ownership of air conditioners in India has tripled since 2010, reaching twenty-four units per 100 households, due to rising heat and increased incomes.[18] The lefthand pie chart in Fig. 5 shows emissions by nations in 2020 resulting from burning fuels. It establishes that China, the United States, and India are responsible for half of all emissions in the world. Although China has had higher emissions in the last two years, historically, the U.S. has the highest share of cumulative emissions, as shown in the righthand Fig. 5 pie chart, making it the highest per capita user of energy. China has about 1.39

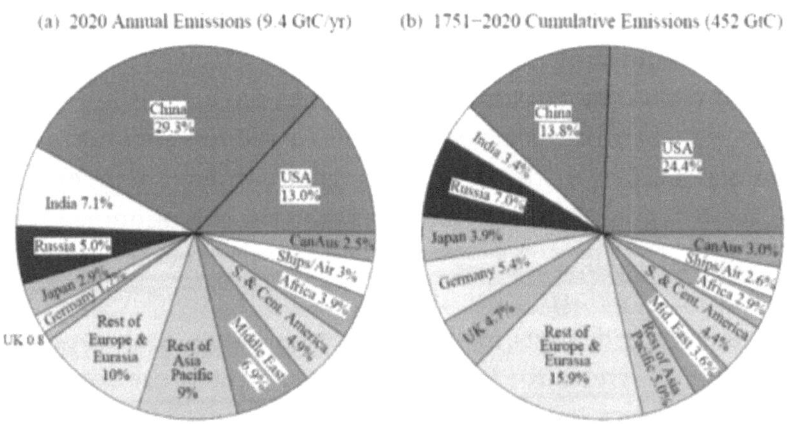

Fig. 5 - Left. Annual Emissions of Nations in 2020. China has surpassed all nations.
Fig. 5 - Right. Cumulative Emissions from 1751 to 2020. The U.S. has the highest emissions.
Credit for Pie Charts: Dr. James Hansen's post on October 26, 2021

billion people, and India has about 1.4 billion. The U.S. population is 327 million, which is less than one-fourth of China or India, and its per capita emissions since the Industrial Revolution are the highest. China and India's cumulative per capita emissions since the Industrial Revolution are significantly lower than the U.S. and other industrialized nations.

HISTORIC RESEARCH ON CARBON DIOXIDE

Joseph Fourier, a French mathematician in 1825, Eunice Foote, an American researcher in 1856, and Joseph Tyndall, an Irish chemist in 1863, separately identified the warming effects of carbon dioxide. In 1896, Svante Arrhenius, a Swedish chemist, was the first to calculate that a doubling of carbon dioxide from pre-industrial levels would cause the Earth to warm by several degrees, with the high northern latitudes likely to warm by twice as much.[19]

Dr. Charles David Keeling, an American scientist, was devoted to measuring atmospheric carbon dioxide starting in 1957. His

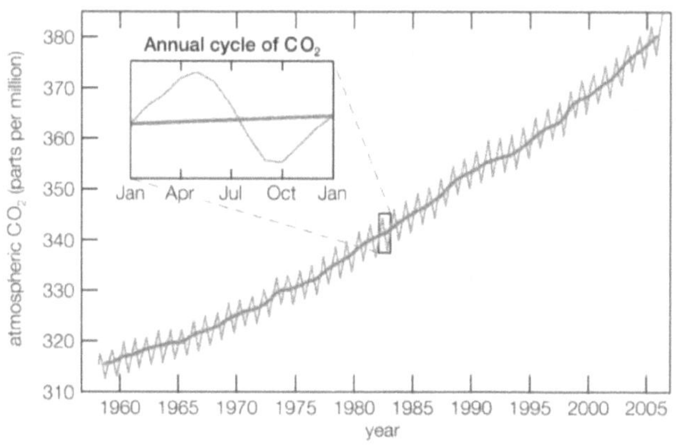

Fig. 6. The Keeling Curve shows an increase of atmospheric CO$_2$ PPM (parts per million) since 1958. Remarkably, the rise in global temperature follows the same curve as seen in Figure 1 in Chapter 1. Graph Credit: Scripps Institution of Oceanography at UC San Diego[21]

graph is called the Keeling Curve. He missed the birth of his first child because he religiously went out every four hours to measure carbon dioxide at Mauna Loa Laboratory in Hawaii. The Keeling Curve revealed annual oscillations and an average carbon dioxide amount that increased every year (Fig. 6). The CO_2 concentration was 315 parts per million (ppm) when he started in 1957, and the estimated global daily CO_2 level on April 3, 2024, was 421.84 ppm, according to NOAA.[20]

Thomas Edison, 1847–1931, raised a potent warning for humanity many years ago: "We are like tenant farmers chopping down the fence around our house for fuel when we should be using nature's inexhaustible sources of energy—sun, wind, and tide. I would put my money on the sun and solar energy. What a source of power! I hope that we don't have to wait until oil and coal run out before we tackle that."[22]

ENERGY CONSUMPTION AND SOURCES

Global dependence on coal, oil, and natural gas is starkly represented in the graph below showing majority of energy coming from these fossil fuel sources while renewable power in minutely

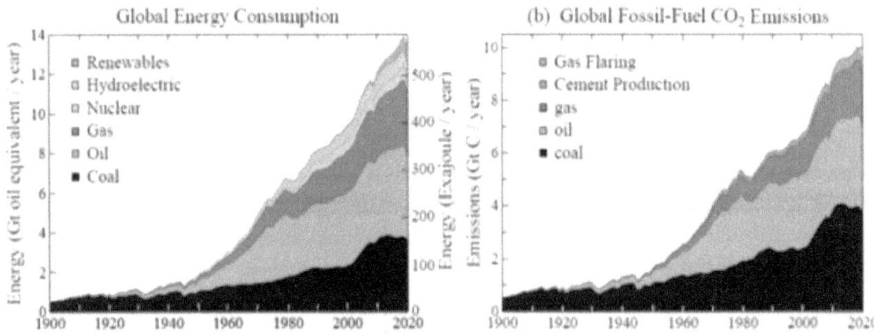

Fig. 7 - **Left.** The energy sources responsible for our energy consumption.
Fig. 7 - **Right.** GHG emissions by energy sources.
Graph Credit: Dr. James Hansen's post on October 26, 2021

small amounts. Energy consumption has shot up since the 1950s and so have the global fossil-fuel-generated CO_2 emissions. Notice that emissions from cement production are so large (eight percent) that they show up on the global emissions graph. Renewable energy forms a small portion of the total energy mix.

Renewables like solar, wind, and hydroelectric energy do not add to emissions, and neither does nuclear energy production. This highlights the change we want to make, a shift to low-emissions energy sources. Scientists are warning us that the path we are on is not sustainable.

What are the sorely needed actions? Sustainable planning of our towns and cities to reduce transportation costs and increase the walkability of the green spaces within our communities. Cement production is responsible for eight percent of all emissions. We must invest in new innovations in building materials to lower the use of cement, steel, aluminum, and glass. Building code improvements are the best way to achieve energy efficiency and create resilient structures. All of these are critical if we have the will to improve building performance by reducing demand for energy.

How do we decrease consumption and increase large amounts of safe and clean energy on a 24/7 basis? A starting point is to demand action from policymakers. When local, state, and federal policymakers listen to professionals—including those in building construction, as well as scientists, researchers, investors, and economists—our chances for transitioning to clean energy sources greatly increase.

INEQUALITIES

India's poor in rural and urban areas use a small number of resources for bare survival. These innocent people are already impacted disproportionately by extreme weather events. They suffer much more from rises in temperature, especially outdoor workers in agriculture and construction and those in India's large informal sectors who make a living by selling products on the streets. The rise in the income levels of the rich in the U.S. has created a huge inequity when it comes to the rise in emissions (Fig. 9). The richest ten percent in the world are responsible for the highest emissions, but the emissions of the richest ten percent in the U.S. are off the chart, as shown below.

According to Somini Sengupta, "Nearly everyone in the United States, even those in the lowest income brackets, produces a lot of

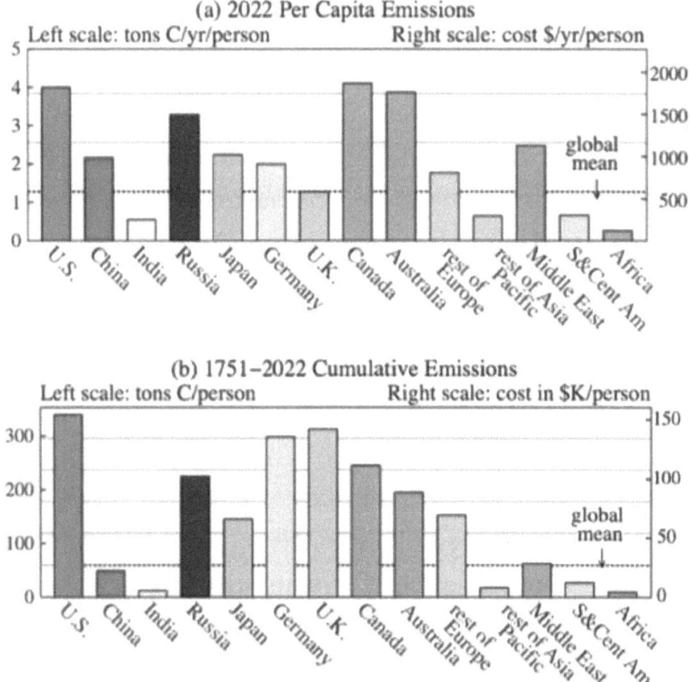

Fig. 8 - **Top** - Per Capita Emissions (tons of carbon/person) in 2022 by countries. The U.S., Canada, and Australia are at the top. Russia remains near the top. India and China's emissions are much lower.

Fig. 8 - **Bottom** - Per Capita Cumulative Emissions from 1751 to 2022. The U.S. is at the top, but it is going down. Germany, the U.K., Canada, and Russia are near the top, but there is a noticeable drop in the U.K.'s and Germany's emissions. China's and India's per capita emissions are much lower compared to industrial nations but they are rising.

Diagram Credit: Dr. James E. Hansen

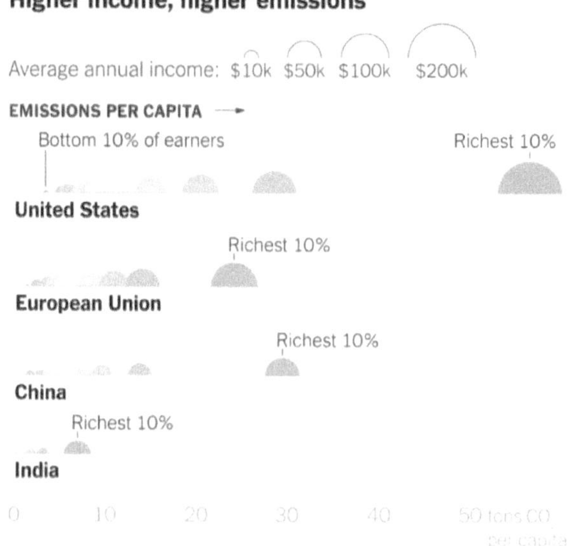

Fig. 9. Emissions by income in the United States, the European Union, China, and India Source: *The New York Times*, "Climate Forward: The American Exception," by Somini Sengupta, February 28, 2023

climate pollution relative to (almost) everyone else in the world. It's the way our economy is built. We take for granted long commutes and frequent flights. Our electricity comes from sources that are relatively carbon-intensive. The rest of the world is different."[23] The per-capita climate pollution in the United States remains much higher in comparison to people in any other part of the world, putting a higher burden on Americans to act and collaborate with nations.

PREDICTIONS — TEXAS 2036

"The latest edition of 'Future Trends of Extreme Weather in Texas' reveals a concerning acceleration in extreme weather conditions across the state," according to the report produced by the State of Texas climatologist at Texas A&M University, Dr. John Neilson-Gammon,[24] who is also a Regents Professor at A&M. This report was prepared in collaboration with Texas 2036 (https://texas2036.org/weather/). This study details significant increases in 100-degree days, intensifying droughts, and heightened urban flooding events. Projections show a dramatic rise in extreme heat, with 100-degree days becoming four times as common in 2036 as they were in the 1970s and 1980s. This will impact built environments, design strategies, and the demand for energy, especially for electricity to run A/C systems.

Texas residents living in older public housing in major cities do not have central A/C in their apartments. A few have installed window units. Most others have no choice but to resort to fans. 100-plus-degree days outside could drive inside temperatures to ninety-plus degrees if the buildings are not energy efficient and do not have cross ventilation. According to the Centers for Disease Control,[25] people of all ages are susceptible to heat-related illnesses, not just the elderly and children. Electric fans help but do not prevent such illnesses. Heat can also exacerbate asthma, as well as heart and lung diseases.

Every summer, people have donation drives to deliver box fans to those who are in dire need. We can do better. City, state, and federal governments must invest in energy-efficient housing to keep the heat out, provide cross ventilation, and help inhabitants access options to cool their homes. Every major city, including Houston, Dallas, San Antonio, and Austin, has programs in place to help the most vulnerable people living in poverty to combat high heat, but these programs alone are not adequate. Here is what Austin Energy and the housing program in Austin are doing.

CITY OF AUSTIN'S HOUSING PROGRAMS

The City of Austin has a free weatherization program for low-income families that was started in the 1980s by Austin Energy, which is a city-owned utility. Despite the downtown skyscrapers popping up for affluent residents, homelessness and affordable housing issues continue to plague Austin. Disparity in our city is growing, and the gentrification of East Austin has forced prices up and longtime residents to move away.

In 2019, the city approved Resolution 20190509-027 to create a program called Affordability Unlocked, which is beginning to generate affordable, sustainable housing projects for low-income Austin residents. The quality housing it enables can take the city on a path to cool its climate in an equitable way. Affordability Unlocked is just a beginning and one of many possible answers for Austin and any city that has the will to safeguard the residents in their patch of the planet.

CHAPTER 7

REMOVING CARBON FROM THE PROCESS OF BUILDING

PASSIVE DESIGN STRATEGIES

How do we expand the skills of our ancestors and the innovations of many bright minds since that time to find comfort in today's buildings by keeping sustainability and climate restoration front and center in the planning and design process? Since the building code and site permitting requirements are given, project teams must make important decisions during the preliminary planning phase:

- Building location: Review site conditions, stormwater drainage, floodplain, tree locations, vegetation, and soil types. There are many lessons we can learn from nature to manage water and vegetation at sites.

- Orientation and wind movement: Understanding the daily and seasonal movement of the sun and wind helps with the location of windows, doors, and covered canopies and verandas for cool light and ventilation. Being observant of these factors will not only save energy and costs but will add to the comfort and quality of life for inhabitants.

- Native trees and vegetation: Plant native vegetation and strategically shade windows in hot climates and get sun inside during cold weather. Native plants could survive through seasonal changes and have lower water consumption needs.

- Roofs: Take into consideration the outsized role that roofs play in all climates, specifically in hot climates. Keeping the heat and the water out of a building should be the major functions of a roof. White coatings and light-colored materials are the best way to reflect heat. Dark asphalt-shingle roofs are quite prevalent in residential developments in the United States. By switching to a light or white roof, A/C use is cut down, which is particularly relevant as the temperatures keep going up.

- Exterior facades: Building enclosure design provides multiple opportunities to provide light and views by shading the openings to protect from hot sun or cold breeze.

- Insulation: Roof and wall insulation to reduce heat gain in summer and heat loss in winter.

- Materials: Using local materials was the only option for builders of the past. They didn't have fossil-fueled transportation. Now, we have capabilities to transport materials from and to faraway locations, adding to emissions from transportation. Learn how materials are manufactured and demand higher transparency from manufacturers to declare the content of the products and their embodied carbon to reduce emissions from their production, transportation, and placement while ensuring the health effects of ingredients used.

- Sustainable planning: While designing communities, find ways to decrease impervious cover over land and increase walkability and green spaces.

PASSIVE CLIMATE DESIGN OF THE TORRENT RESEARCH CENTER IN AHMEDABAD

The Torrent Research Center is a skillfully designed complex of laboratories and offices that uses passive design strategies to create comfort in the hot and arid climate of Ahmedabad. It is one of the largest projects in Asia to use building design and ventilation for cooling.

The project team[1] that designed and built the research center used a passive downdraft evaporative cooling method to lower the inside temperatures by twelve to thirteen degrees Celsius (thirty to thirty-two degrees Fahrenheit). They used a combination of ceiling and exhaust fans, which operated according to season: hot summer, hot and humid monsoon season, or winter. Shading the windows on both the horizontal and vertical planes reduced solar heat gain. The team had the insulated walls painted white and the insulated roofs finished with mosaic tiles to keep the heat out. The use of light color and reflective mosaic

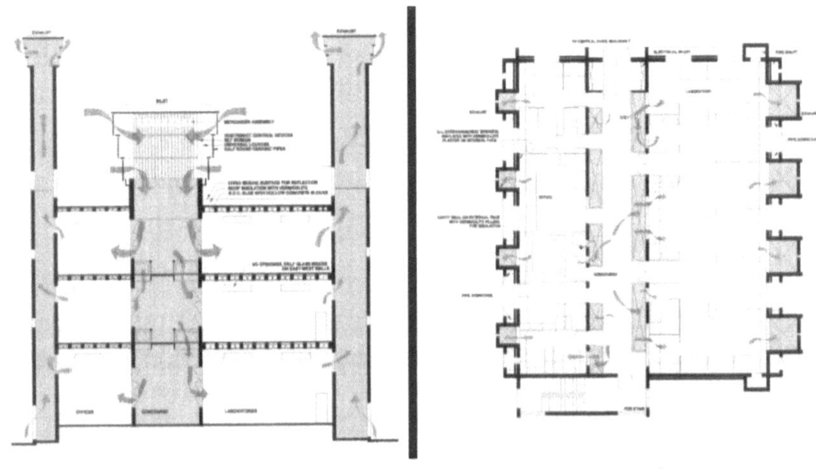

Fig. 1. Section of the Torrent Research Center on the left and a partial floor plan on the right. Designed by the architectural firm *Abhikram* in Ahmedabad, India, and Environmental Consultants Short + Ford Associates, London, U.K.[1]

Drawing Credit: Parul Zaveri of *Abhikram* and the project team tiles is a lesson from the past, as shown in Chapter 2.

This unique building was completed in 2000. Since then, the research center has conducted reviews and user surveys to show the building's continued comfort level and higher levels of staff productivity. The owners of the center saved money after paying initial expenses and reduced emissions at the same time.

EMBODIED CARBON OF BUILDING MATERIALS

What is embodied carbon? It is carbon that is emitted over the life cycle of building materials. Concrete, steel, aluminum, and glass are widely used materials with large carbon footprints. Embodied carbon is generated in the extracting, manufacturing, transporting, and installing of these building materials for construction prior to project completion. Carbon is also emitted during operation, maintenance, and disposal at the end of its life cycle. In the building design and construction process, a reduction of embodied carbon is the most effective way to address global warming, reduce future emissions, and cool our environment.

CONCRETE

Concrete is so widely used in construction that it is hard to imagine any project, especially projects like high-rise skyscrapers, parking garages, bridges, dams, tunnels, and sidewalks without concrete. A comparable strong and durable material is difficult to find. The Pantheon in Rome was the first major structure that had a concrete dome, which spans 141.7 feet in diameter, and it is still standing (worldhistory.org/Pantheon/).

Cement is a major ingredient in concrete that, using water, binds all aggregates together. Cement is also a major contributor

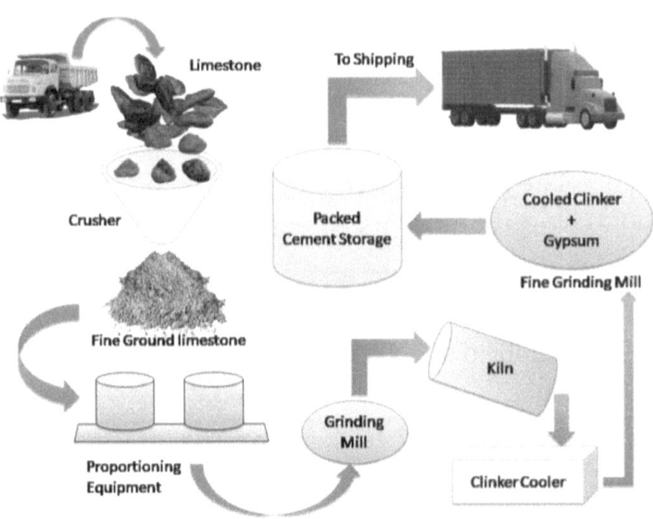

Fig. 2. Flow diagram of cement manufacturing from extraction from the mine to shipping; cement must be cooked at temperatures of 2,700 degrees Fahrenheit, according to the Portland Cement Association.[3] Credit for Diagram: The Machine Design, machinedesign.com

to climate change because of the energy required in its manufacturing and transportation. Cement's carbon emissions make up eight percent of the total global emissions, according to the think tank Chatham House.[2] It shows up on the global emissions graph along with coal, oil, and natural gas (see Fig. 8, Chapter 6).

In most of Austin's new building projects, the foundation structure uses concrete beams and slabs. The pedestrian bridge project over Town Lake has pile foundations that were driven deep into the ground. When buildings are built in clayey soil in certain parts of the Austin area, deep pile foundations are a part of the design elements. While thinking about concrete driveways, parking lots, and sidewalks, we need to think of other options. The Austin Animal Center project used pervious pavers for the parking lot and sidewalks, allowing water to pass through to the ground underneath. This reduced cement use and helped with stormwater drainage on a flat site.

The reduction of cement in concrete has been an issue around the world for a long time. Research projects are ongoing to find substitutes for cement and other ingredients of concrete mix. Common materials used to make cement include limestone, clay, shells, chalk, shale, blast furnace slag, silica sand, and iron ore. In a cement manufacturing plant, these ingredients are heated at extremely high temperatures, and then the rock-like substance is ground into the powder called cement (Fig. 2). This highly controlled chemical process requires high amounts of energy to manufacture, with added energy for transportation afterward.

Cement manufacturers are looking for ways to reduce their carbon footprints. Fly ash, a byproduct of a coal power plant, is used when available to partially replace cement, thereby reducing the amount of cement, its embodied carbon, and the cost of concrete. But as coal power plants are closing, we need to find new replacement materials.

For City of Austin projects, standard specifications for concrete-mix design have options to reduce a certain amount of cement content by replacing it with fly ash, blast-furnace slag, or other approved materials. Government agencies around the world are trying to find alternatives to cement. The City Council of Austin recently passed Resolution 20230420-024 (see Appendix 3), requesting staff to present alternatives to revise concrete-mix specifications. The process of identifying and approving revisions to concrete specifications is slow, but climate urgency requires that we find solutions expeditiously. In my current City of Austin project, the Dove Springs Public Health Center,[4] we have used a concrete mix that replaces a small amount of cement with lime. What are the positive impacts of replacing cement with lime? This project uses 1,300 metric tons (1,730 cubic yards) of concrete. By replacing part of the Portland cement with lime, we keep an equivalent of 13,886 Kg. CO_2 from going into our atmosphere. It is a drop in a bucket, but a bucket gets full with many drops!

For the last twenty years, the city of Zurich, Switzerland,

has championed the use of recycled concrete in buildings to preserve nature and reduce local urban waste. By requiring a minimum share of recycled aggregates in concrete for new buildings, Zurich's public procurement policy created a market for recycled concrete in construction. It introduced a requirement in 2005 that all publicly owned buildings must use recycled concrete. This was a start. Now, around ninety percent of the concrete used in public buildings is recycled concrete, consisting of up to fifty percent of recycled aggregates in Zurich.[5]

When a public entity sets an example through its own buildings, the markets get signals to start transforming the production of cement and mix-design options for concrete. An international group of researchers, including the Rocky Mountain Institute (RMI), have been working on a concrete-mix design using a low-carbon cement called Limestone Calcined Clay Cement (LC3).[6] Clay is available in viable quantities, and LC3 seems very promising. India recently passed a new code to allow the use of LC3 in concrete with comprehensive guidelines for production and testing. Researchers are cautiously optimistic.

Blue Planet™ is working on capturing carbon and turning that into valuable building materials like aggregates used in concrete, storing it permanently in concrete without any leaks into the atmosphere.[7] Aggregates form the largest component of concrete and if this innovation becomes successful, it will not only cut down emissions of concrete but sequester carbon from the atmosphere to help restore our climate.

Sublime Systems is a leader in creating carbon-free cement. "It produces a cement at room temperature that can drop into the existing ready mix supply chain and by complying with the American Society for Testing and Materials standards" (cnbc.com/2023/06/24/sublime-systems-co-founders-making-the-electric-vehicle-of-cement.html). Sublime Systems completed a pilot plant at the end of 2022. The next step is to go from the 100-ton pilot plant to a 30,000-ton-per-year demonstration plant.

STRUCTURAL STEEL

Steel is another structural material that is used in large quantities in building and infrastructure projects. Like concrete, steel production is one of the most energy-consuming and CO_2-emitting industrial activities in the world. The main ingredient of steel is iron, mined from the earth. Molten iron from the blast furnace is sent to a basic oxide furnace, which is used for the final refinement of the iron into steel.

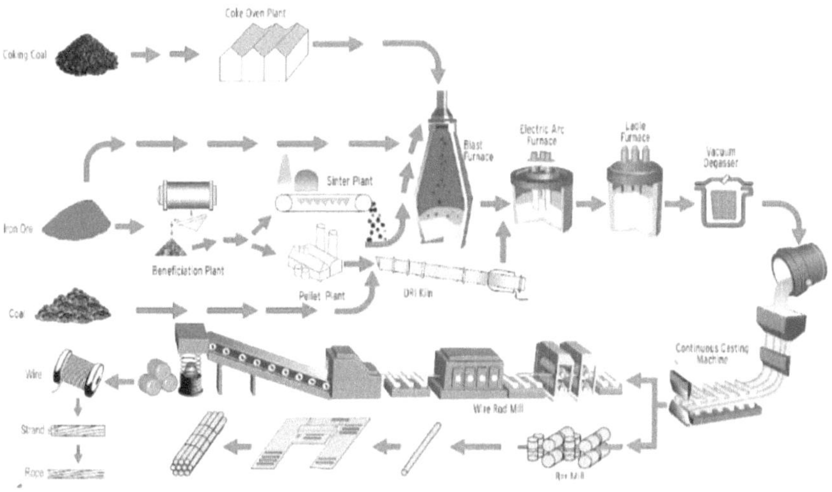

Fig. 3. Steel Bar Manufacturing Process.
Diagram Credit: https://www.singhaniainternational.com/products-details/wire-rod-coils/126, by AmiEffectives-21.46

High-purity oxygen is blown into the furnace, which combusts the carbon and silicon in the molten iron. The basic oxide furnace is fed with fluxes to remove any final impurities. Alloy materials may be added to enhance the characteristics of the steel (mechanicstips.blogspot.com/2016/03/steel-making-processing.html). The type of energy used in making steel—coal, natural gas, or electricity—determines its CO_2 emissions. Researchers are working on options to lower the use of fossil fuels to reduce carbon emissions resulting from steel production. Once manufactured at a plant, the metal is transported to the factories making structural steel members, before the structural steel is transported to construction sites for installation.

ALUMINUM IN BUILDINGS

Bauxite is a sedimentary rock with relatively high aluminum content used for manufacturing aluminum. Global aluminum production has been rising due to population and economic growth. But aluminum plays a major role in a clean-energy future. It is lightweight and flexible and used heavily in building skyscrapers. It has multiple uses, including making solar panels. According to the International Energy Agency, emissions reduction today from steel and aluminum are challenging due to heavy reliance on fossil fuels, process emissions from incumbent routes, and high-trade exposure. Increased scrap recycling and mass deployment of innovative technologies, including carbon capture, are key levers for reducing emissions for both steel and aluminum.[8]

GLASS IN BUILDINGS

The introduction of glass in buildings was a significant life-changing experience for interior spaces. Glass was difficult to make, and that is why it remained a luxury item. Prior to large quantities of glass production, England had two taxes: the window tax and glass tax, which was also called "a tax on air and light" and was eventually abolished.[9] But for some time, due to taxes, people bricked up window openings, making dark and airless rooms.

Because glass is made in furnaces that use extremely high temperatures, glass manufacturing has a high carbon footprint.

As of 2022, the global carbon dioxide emissions from glass production amounted to ninety-five million metric tons. The total glass production in the European Union (EU-28) increased from 31.6 million metric tons in 2013 to a high of 39.5 million metric tons in 2022.[10]

Glass manufacturing is an energy-intensive industry that is mainly fueled by natural gas. Glass is made by heating limestone, sand, and soda ash to 1,500 °C. This heat comes from natural gas, and it accounts for between seventy-five and eighty-five percent of the carbon emissions from glass manufacturing.[11] Glass is an essential material in buildings. To reduce its embodied carbon, we need to start replacing heating sources in glass manufacturing with cleaner energy and recycling glass during demolition.

DESIGN CONSIDERATIONS FOR CONCRETE AND STEEL

Rocky Mountain Institute (RMI) is a nonprofit and nonpartisan organization that is working on global energy systems. It reports in its publication, *Profitably Decarbonizing Heavy Transport and Industrial Heat*, that concrete and steel savings can be achieved profitably through better structural design. It further points out that such engineering design methods had reduced New York's Freedom Tower concrete use by forty percent and China's Shanghai Tower by twenty-four percent.[12] The Indeed Tower in downtown Austin is a LEED Platinum building, which had a thirty-five percent reduction in concrete by using post-tension slab construction.[13]

WOOD AND MASS TIMBER

In Europe and Canada, the use of mass timber construction is increasing in both low-height buildings and in buildings taller than ten stories. The design and construction industries are beginning to embrace mass timber in the United States. Mass timber is an engineered wood product that can sequester carbon for years after it is installed. It is engineered for high strength like concrete and steel, but it is lighter than them.[14] Substantial improvements are made to meet the fire code and moisture management of exterior facades during construction.

This new product seems to have great potential, and like all other building products, researchers should responsibly examine and study its embodied carbon from cradle to grave and its potential for carbon neutrality in applicable projects. It also needs to ensure that it results in no negative impacts on forests, the lungs of our planet that absorb harmful carbon-based pollutants.

What may be appropriate in the Northwestern region of the United States, in Canada, and in Austria may not be applicable in the U.S. Southwest, in parts of India, or in other countries where local materials present different challenges. Professionals who are consciously planning and designing to reduce the embodied carbon content of building materials have explored wood construction where appropriate and reduced the use of concrete and steel and, therefore, carbon emissions.

To keep sustainability and climate restoration front and center, designers and builders must be prudent and embrace all options that are locally viable. They have the power to reduce emissions that result from building design, construction, and operations. Some zoning and building code requirements mandate certain planning decisions long before a project becomes a reality. Sustainable Development Code[15] is a free resource for governments and communities of all sizes and budgets. It supports more resilient, environmentally conscious, economically secure, and socially equitable communities.

Today, every building project can and should use recycled materials to the maximum extent possible. Besides concrete, steel, aluminum, and glass, it is crucial to invest in innovations that reduce the carbon footprint of all building materials, like insulation, gypsum board, acoustical tiles, air barriers, roofing, building enclosure, and others. It is a long list.

AMERICAN INSTITUTE OF ARCHITECTS

The American Institute of Architects (AIA) is the largest professional organization for designers and architects. As an organization, it did not officially come out to recognize the climate crisis until early 2006. However, many AIA members have been involved in designing sustainable buildings using passive climate strategies for a long time, and they are major participants in the USGBC's activities. Architects, designers, engineers, planners, material manufacturers, and contractors give presentations at the USGBC conferences. The City of Austin and other local governments around the world are placing a big emphasis on reducing carbon emissions in buildings through policy and code changes.

In 2006, Edward Mazria of the AIA created the 2030 Challenge.[16] His book *The Passive Solar Energy Book* is an outstanding resource for designers who want to learn about passive design strategies. He has been tirelessly working to decarbonize built environments by redefining their planning, design, and construction processes. The goal of this commitment was for all new buildings and major renovations to be carbon-neutral by 2030. *Carbon-neutral buildings*[17] are defined as buildings in which the energy generated at the project site is equivalent to the energy needs of the facility. Mazria's book allows architects, engineers, designers, and builders to join a growing movement dedicated to addressing climate change that can be demonstrated in concrete and verifiable ways.

As part of the 2030 Challenge, Edward Mazria worked on

the China Accord, which Chinese and American firms signed in October 2015. It was a commitment to plan and design cities, towns, developments, and buildings in China to low-carbon and carbon-neutral standards. Mazria wrote, "We understand our moral and professional responsibility to address the issue of greenhouse gas emissions if we are to stay within the two-degree-Celsius threshold established by the international scientific community, and the Accord is just the beginning of our joint efforts. We have a long and exciting road ahead of us to decarbonize the built environment."[18]

Architects and engineers who specify building materials face a conundrum. In the global economy, professionals have little or no direct control over the process by which materials are manufactured and transported. The AIA is just beginning to track the embodied carbon of materials used in buildings. Even sustainably designed buildings could contain a large amount of embodied carbon when their carbon footprint is fully accounted for. Professionals are digging deep into the Environmental Product Declarations, or EPDs, to make products more transparent in terms of embodied carbon and the health effects of chemicals used. The AIA recognizes the need for policy changes by governments, specially building code requirements, to meet the 2030 Challenge.

BUILDING CODE REQUIREMENTS

Examples can support how places with strict code requirements have withstood extreme weather events compared to places without such codes. When Hurricane Ian hit Florida in 2022, its Babcock community showed how to build in the pathway of hurricanes and still protect the inhabitants and their properties. The international and national codes must be adopted by the State. Powerful corporations in the U.S. and around the world have influenced federal agencies, states, and local governments to forego adopting codes that may affect their bottom lines.

Carbon Architecture[19] is a design movement that replaces raw materials used in building construction with bio-based materials that sequester carbon. These materials include wood, clay, bamboo, straw, and hemp. Designing carbon-neutral buildings combined with passive climate-design principles is a very effective step forward to increase sustainability and reduce climate pollution.

Advocacy by architects, engineers, and other building professionals at the local, state, and national levels can have a huge impact on policymakers. These are the experts who know how to transition our communities away from fossil fuels in favor of low-carbon technologies. This collective support from industry and government would not only make it easier to reach the AIA's 2030 Challenge, but it would improve the quality of life for all.

An example of Hong Kong's Bauhinia Bud Building has lessons for architects and building professionals.

LESSONS FROM HONG KONG'S BAUHINIA BUD BUILDING

Dame Zaha Hadid was a prominent architect of the twentieth and twenty-first centuries who passed away in 2016. Born in Baghdad, she practiced in Britain and was one of the most creative architects of her time. Her architectural firm is still called Zaha Hadid Architects, and its team worked on a project in Hong Kong that transformed a thirty-six-story existing car park into an urban oasis adjacent to the Charter Garden in Central Hong Kong. Construction was completed in 2022. It is a beautiful, sustainable building that has achieved LEED Platinum and WELL Platinum certifications, which is currently the highest level of certification achievable in the industry for those rating systems.[20]

This region gets hit with powerful summer typhoons. The building facade is composed of four-ply double-laminated, double-curved insulated glass units, the first of their kind used in Hong Kong, to effectively insulate the building and reduce its cooling loads as well as build resilience to the extreme climate. Weather stations are installed to monitor real-time outdoor conditions, including pollutants, ozone, daylight, wind speed, rainfall, temperature, and humidity. It is designed to achieve a twenty-six-percent reduction in electricity by using smart chillers, efficient HVAC equipment, and daylight sensors.

Powerful typhoons in Hong Kong and Southeast Asia are becoming increasingly frequent and are hitting the area with higher intensity due to climate change. The owners of the Bauhinia Bud building responded to the changing design conditions by spending enormous sums of money in manufacturing a special glass requiring high amounts of energy and resultant emissions to clad the exterior of this building to withstand those forces.

Billions of people around the world are already victimized by extreme heat waves, flooding, droughts, sea level rise, wildfires, and air pollution due to the intensifying extreme weather events. They contribute little to the climate crisis but are helpless when weather disasters strike. So, how do we design sustainable modern buildings for common people using low-cost strategies?

SUSTAINABLE DESIGN STRATEGIES FOR MODERN BUILDINGS

All passive strategies described earlier—building location, orientation and wind movement, native trees and vegetation, roofs, exterior facades, insulation, materials, and sustainable planning—

apply to modern buildings. These are the first steps in reducing environmental damage from the impact of building structures. These steps would help reduce heating and cooling loads and minimize the impact of building materials on the environment. With technological developments, modern buildings have additional strategies to find comfort by using mechanical systems, artificial lighting, electricity generation, smart controls, carbon capture, and modern equipment and devices. All of them add substantial amounts of electricity, and it is up to us to design in ways that minimize the use of electricity.

- Building insulation: Insulation is the most important design feature to keep the heat from entering the building in hot climates and preserving the heat inside in cold climates. Roof insulation is especially effective in hot climates since a large portion of heat comes from the roof. For insulation to work, weatherization is crucial. Electric utilities like Austin Energy can help with weatherization inspection and rebate programs.[21]

- Reflective roof and wall surfaces: White or light-colored reflective surfaces reduce cooling load and air-conditioning, which greatly contributes to a building's emissions in hot climates. Roof and wall gardens absorb pollutants, create cooler environments, and reduce emissions.

- Shading devices and energy modeling: Coordinated overhangs and canopies help protect openings from getting direct sunlight and reduce cooling load and air- conditioning. With computerized design processes and energy modeling, designers can study shade patterns to coordinate overhangs. Overhangs also help with water penetration inside, protecting building materials and keeping moisture out.

- Heating, ventilation, air-conditioning, and cooling (HVAC) equipment: HVAC is a major contributor to building emissions once the building is occupied, and it continues to add emissions for years. The first step is to reduce heating and cooling loads through passive design strategies. Other steps include the preparation of an energy model to determine needs, one that avoids oversizing HVAC equipment. Using heat pumps improves efficiency and switching to electrification lowers emissions if electricity supply is coming from clean energy sources.

- Commissioning HVAC and other equipment: Commissioning ensures the coordination of design and construction documents, review of energy modeling, and proper installation and functioning of all systems for a desired performance for users of buildings.

- Continuous air barriers: Air barriers protect cool and hot air leaks, saving energy. They also protect building materials from water penetrations during rain events.

- Solar panels: Install solar panels with battery packs to meet most of your needs for electricity. Make all buildings solar-ready at the time of construction. Many cities, including Austin, have this code requirement.

- Rainwater collection: Rainwater collection reduces treated water coming from water treatment plants, which pump water to distant locations using energy. Saving water saves energy.

- Reclaimed water: Austin Water (the city's water department) began providing reclaimed water in 1974. Reclaimed water is recycled from wastewater generated by homes and businesses and treated for non-potable uses. Reclaimed water is less expensive to treat and costs almost half as much as potable water.[22] There are many cities in the U.S. that provide reclaimed water for landscapes, lawns, and chillers. But in most cities, treated potable water is used for landscape and lawns. The U.S. Environmental Protection Agency offers more information about reclaimed water at epa.gov/ waterreuse.

- Rain gardens and permeable paving: This helps to lower stormwater runoff from buildings, eliminating the need to build concrete water-quality ponds and stormwater infrastructure.

MODERN CLIMATE DESIGN OF THE AUSTIN CENTRAL LIBRARY

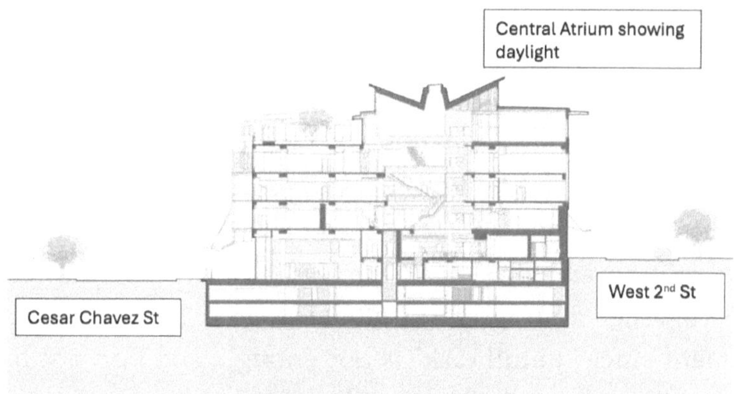

Fig. 4. A building section of the Austin Central Library. It has used many strategies to achieve a LEED Platinum certification, including abundant daylight without adding heat load, shading devices, solar panels, effective use of local materials, and other measures that save water and energy. Lake Flato Architects of San Antonio, Texas, was the prime consultant, with subconsultants, on this project.[23]
Drawing Credit: Prime Consultant Lake Flato Architects

These are just a few strategies for passive design and design with modern technology. There are hundreds of thousands of creative and innovative planners, designers, and builders around the world who have used solutions to construct built environments for the protection of people in this warming world. Austin, being an early leader, has consultants who are committed to designing and building sustainably. There are other cities that are leading the way in adopting building codes that would contribute to lowering carbon. Books and articles combining passive strategies and modern technologies to move toward decarbonizing buildings are available. But we are still far from there.

GREEN BUILDINGS ARE ABSOLUTELY CRUCIAL AND OTHER ACTIONS

All sustainable building strategies and building code improvements help us to transition toward *net-zero* buildings[24] by lowering future emissions.

But climate science is clear: the carbon-based and greenhouse gas emissions we have been spewing into the atmosphere for years will continue to warm the planet, making it harder to decarbonize buildings unless we create a political will for our lawmakers to enact doable policies to protect us. Even if we have a constitutional right to vote for and elect leaders who are willing, it will be an uphill task for leaders from all countries to come together and agree on global action. That is an understatement since "uphill task" is not enough, and I do not have a phrase to describe the gigantic effort needed.

That means that we must double down on our advocacy work for the sake of literally everyone on the planet.

CHAPTER 8
WHY IS ADVOCACY IMPORTANT?

WHY ADVOCATE?

We seem to have a disconnect between what most Americans want and the lack of adequate measures by lawmakers to address climate change.[1] To end the disconnect, all of us need to first understand what is happening to our environment, how it is wrecking lives and livelihoods around the globe, and how policies that will restore our climate can become a priority. A few extreme weather events (Chapter 1) gave you a glimpse of possible devastating effects. With increased warnings by scientists and slow actions by global leaders, all of us must step up our efforts.

The Intergovernmental Panel on Climate Change (IPCC) is an international scientific authority on all things related to global warming. It aims to inform policymakers around the world of the risks of human-caused climate change. IPCC has issued six assessment reports since 1988, forecasting climate conditions and actions we need to take to stabilize our climate.

In 2015, at the Conference of Parties (COP21) meeting of nations, the Paris Climate Agreement[2] was signed by almost all nations. They agreed on climate mitigation, adaptation, and finance to keep the global temperature rise below 1.5 degrees Celsius by 2050, as recommended by IPCC. The United States did not sign on to this agreement until 2021.

In November 2021, in Glasgow, Scotland, at COP's twenty-sixth meeting (COP26), nations of the world announced their action plans to reach net-zero emissions to help keep the global temperature rise below 1.5 degrees Celsius by 2050. That optimism soon vanished when coal and other fossil fuels made a comeback, and emissions trajectory turned up again. Since then, there have been two more meetings, COP27 in Egypt and COP28 in Dubai, with the participation of about 200 countries who continue to make pledges to lower the use of fossil fuels, reduce emissions, and keep the global temperature rise below 1.5 degrees Celsius by 2050. But the global temperatures continue to rise. Actions by industrialized nations with high per capita emissions are crucial to help people of poor nations who are victims of extreme weather but are least responsible for causing the climate crisis.

LEGISLATIVE ACTIONS

Two legislative actions were being considered by countries around the world while addressing climate change and rising emissions: 1: carbon fee and dividend (CF&D), and 2: cap and trade (emissions trading schemes) to reduce emissions. With cap and trade, total greenhouse gas emissions are capped, and companies must purchase permits for every ton of gas emitted. The more expensive the permit, the greater the incentive to reduce emissions. Cap-and-trade action was the centerpiece of the last major effort to price carbon in 2010 in the U.S. It ultimately failed due to opposition from businesses, most Republicans, and some Democrats.

Dr. James Hansen[3] was the first scientist to testify before the U.S. Congress regarding climate change concerns in 1988, and he strongly supported the carbon fee and dividend approach in his book *Storms of My Grandchildren*. In this method, a fee is collected at the mine, well, or port of entry for each fossil fuel (coal, oil, or gas) at its first sale in the country. The fee to be collected was determined based on the amount of carbon dioxide in the fuel. So, if the U.S. Congress passes the Carbon Fee and Dividend Bill[4] at some point, they would determine the dollar value based on damages done by a ton of carbon dioxide. The public wouldn't pay any fee directly but would end up paying an increase in the price of goods, as companies are likely to pass the fees they are charged on to customers. Customers would pay increases in the price of goods from the dividends they receive every month. This is a strong incentive to buy cleaner products that can drive markets away from coal, oil, and gas.

Fuels such as gasoline or heating oil, along with electricity made from coal, oil, or gas, are affected directly by the carbon fee, which is set to increase every year. The carbon fee will rise gradually so that the public will have time to adjust. Lifestyle choices, including type of vehicle, energy efficiency improvements in homes, the use of Energy Star appliances, and the use of efficient air-conditioning systems, could lower an individual's carbon footprint. The idea is that companies affected by the carbon fee on their products would find ways to transition to cleaner fuels and cleaner products, and public demand for cleaner products would keep increasing. This would send a strong message to the markets.

Dr. Hansen recommended that, under this fee-and-dividend action, 100 percent of fees collected from fossil fuel companies at the mine or well should be distributed to American households. As a result, those who do better than average in reducing their carbon footprint would receive more in the dividend than they would pay in the added costs of the products they buy. This straightforward approach does not require a large bureaucracy. The proposal for a carbon fee and dividend has the support of a vast majority of prominent Nobel-prize-winning economists and environmentalists.[5]

FOUNDER OF THE CITIZENS' CLIMATE LOBBY

Citizens' Climate Lobby's (CCLcitizensclimatelobby.org) founder, Marshall Saunders,[6] was part of an advocacy group called RESULTS that worked to alleviate hunger and poverty around the world. The Al Gore documentary *An Inconvenient Truth* on the climate crisis alarmed him. Seeing that his efforts to help people out of poverty would be in vain if climate change made their homes unlivable, he began talking about the climate crisis.

Saunders thought policymakers would handle the problem of climate change. After giving talks on climate change, he realized that Washington policymakers, who were handing out billions of dollars to the fossil fuel industry, dwarfed personal choices. He realized that people needed to empower themselves, reclaim their democracy, and engage their members of Congress to act on climate change. CCL is a nonprofit, nonpartisan, grassroots advocacy organization focused on national policies to address climate change.

VOLUNTEER WORK AT CCL

I started volunteering with CCL's Austin chapter. This included monthly potluck meetings held in the homes of two gracious Austin chapter founders, Anna Graybeal and Susan Adams. Our chapter and others around the country joined a national call with Mark Reynolds, the executive director of CCL at the time. Today, CCL has more than 456 chapters in the U.S. and roughly 210,000 members in the United States. CCL's international chapters work to influence their governments to adopt similar policies to make a global-level impact.

Before my involvement with CCL, it was not easy to even think of writing letters to the editor. Now, it is a regular practice for me. It is a respectful, non-partisan, and persistent engagement strategy of CCL, which has prepared its volunteers to publish letters and op-eds. The CCL provides guidance and support for writing to newspapers. Members of Congress pay attention to the views of their constituents when published, and publishing letters creates awareness among readers. In 2023, 1,304 letters to the editor and 440 op-eds were published by the CCL volunteers in the U.S., supporting several bills that were signed into law by the U.S. Congress.

CCL volunteers represent every congressional district in the United States. A team of constituents works along with a liaison from the office of the elected member of the U.S. House in each district. Republican Bill Flores represented my congressional district, TX-17, and I was appointed to work with his office. One would think that Austin, with its population close to one million, would have two congressional districts since each district has about 600,000 people, but the Texas Legislature was able to divide Austin into five congressional districts in 2016. Each district started in Austin and then fanned out to other cities and rural areas in every direction.

This type of gerrymandering by either party, which has

Fig. 1. Previous map of Texas District 17 (through 2022).
Source: State of Texas Website (Texas.gov)

long-term negative effects on local populations, is against all democratic values the United States represents. Both parties around the country engaged in gerrymandering after the census data from 2020 was released. When new district maps were drawn, Texas gained two new seats in the U.S. House. Until 2022, I lived in District 17, which consisted of Bryan, College Station, Waco, and rural areas. Austin was a very small part of it. My new representative is Congressman Lloyd Doggett, a Democrat.

LOBBYING IN WASHINGTON, D.C.

The main goal of the Citizens' Climate Lobby (CCL) members is to reach out to our representatives in the U.S. House and Senate, as per Marshall Saunders's original vision. The CCL keeps its ultimate focus on having a federal policy for a carbon fee and dividend (CF&D) policy enacted into law. But each year, there are supportive bills that are important to reach the goal of enacting a carbon fee and dividend policy. Every year in June, CCL volunteers gather in Washington, D.C., for a national conference weekend full of workshops and presentations on understanding the most current developments that would affect the environment and people. The last day is reserved for congressional lobbying in the nation's capital.

Participants vary in age, from high schoolers to senior citizens. In my lobbying session in June of 2024, our youngest member was a seventeen-year-old student from Lubbock, Texas. Most CCL members are volunteers, including highly qualified professionals in different fields. As a lobbying organization, CCL has an organized process to prepare its volunteers to meet with and talk to their congressional members about the CF&D policy and other relevant bills.

Walking the hallways of the Capitol and meeting Congress members in their offices is a daunting task. The offices I visited were small. When no conference space was available, we had standing meetings in the hallway outside the office. My colleagues and I had two such meetings with Rep. Flores's office staff. We were given thirty to forty minutes to present. I had read about the CF&D policy, but now I was trying to articulate it in front of my representative. We met with Rep. Flores's energy aide, Eric Gustafson, who was very knowledgeable on issues related to energy and asked pertinent questions. In 2019, we were finally invited inside Rep. Flores's office and had a very good meeting with Mr. Gustafson.

Representative Bill Flores had his main office in Bryan, Texas, and other offices in Austin and Waco. During the Congressional recess, Rep. Flores spent time in Bryan, his hometown. All my meetings with him were held in Bryan, along with my constituent colleagues from Bryan and College Station. Austin constituents drove a minimum of two hours to reach his office.

CCL'S LIAISON TO LAWMAKERS

Before becoming a congressman, Mr. Flores worked in the oil and gas industry. He was courteous and pleasant and gave us more time than scheduled. He listened carefully to the CF&D policy that CCL supported. He raised questions about how the carbon tax could impact poor working people. We shared our household impact study (Fig. 2). CCL volunteers collected questions from all representatives and followed up with answers or did studies to respond to them. A general response from the members of Congress has been quite positive to CCL volunteers, who conduct themselves in a professional and respectful way.

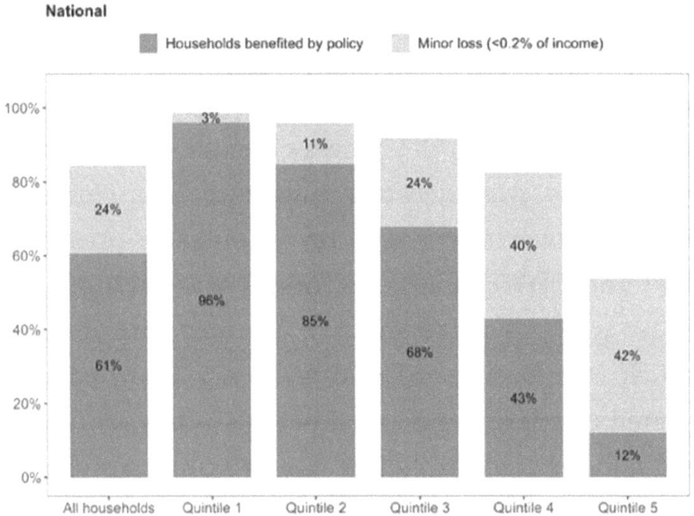

Fig. 2. Percent of households whose carbon dividends exceed carbon costs, ranked by consumption quintile in the U.S. The lighter gray denotes a minor loss, defined as less than 0.2 percent of income (e.g., for a $50,000 income, less than $100 per year). Study Highlights: sixty-one percent of households and sixty-eight percent of individuals in the U.S. end up receiving more than enough in monthly carbon dividends to offset their increased costs. The figure above shows how those net benefits break down across quintiles (each quintile = 1/5 of the U.S. population) ranked by household spending (consumption).
Source: citizensclimatelobby.org/household-impact-study/

ADVOCACY IN BRYAN, TEXAS

I was a liaison to Rep. Flores until he retired in 2020. Like all CCL volunteers, a primary ask of mine was to get my representative to sponsor the Carbon Fee and Dividend policy bill, H.R. 2307. Rep. Flores didn't sponsor the bill. In 2023, the Energy Innovation and Carbon Dividend Act (now H.R.5744), a bill that closely aligns with CCL's carbon-pricing principles, was reintroduced in the U. S. Congress.

It used to be that members of Congress held in-person town hall meetings in their districts. Bill Flores was proud of it, and he did that on a regular basis. I went to a town hall meeting in

Fig. 3. In June 2019, I joined CCL volunteers from around the world on the U.S. Capitol steps in Washington, D.C., to ask congressional leaders to enact national policies to address the climate crisis.
Source: citizensclimatelobby.org

a community center in Bryan, Texas, and realized that none of my constituent colleagues from the Bryan-College Station area could attend. After the presentation, he had a question-and-answer session. We had to submit questions ahead of time, which I did, asking about climate change and his views about actions we should take.

The audience was mostly partisan, with few people of color. I was a minority in that audience. Rep. Flores responded from the stage in the community hall and announced for the first time, "I am not a climate denier." He told the audience that he has solar panels on his house, which implied his support for renewable energy. It was a bold statement at the time, when former President Trump, whom Representative Flores supported, called climate change "a hoax."

During my visits to Bryan, Rep. Flores had shown interest in innovations in battery storage for solar power and carbon-capture technologies, both important in addressing climate change. But it was too much of a deviation for him to transition away from fossil fuels. The best I could offer was my continued engagement. His support for the Carbon Fee and Dividend policy wasn't there.

In-person town hall meetings were becoming disruptive and violent for some members around the country in 2018. Rep. Flores, who was proud of holding face-to-face town hall meetings, had to change to virtual meetings, one for Austin and one for College Station. He regularly held such meetings and invited constituents. He also had a poll at the start of every meeting to decide which topics were important to his constituents. My constituent colleagues and I joined those virtual meetings to ensure that climate change was discussed in every meeting.

Congressman Bill Flores had an education in business and worked in the energy industry for thirty years prior to getting elected to Congress. He served on the powerful House Energy and Commerce Committee. He was interested in government policy that would create jobs, especially in the oil and gas industries. Climate scientists from Texas A&M University presented scientific facts about climate change and called on the government to act. Rep. Flores showed interest in nuclear power.

In fact, my husband, Swadesh Mahajan, and his UT colleagues had developed a design for a fusion-fission hybrid that combines the two nuclear channels of fission and fusion. The hybrid reactor was made possible by one of their inventions, called the SuperX divertor.[7] The hybrid could be used to turn most of the nuclear waste into energy, a kind of greening of nuclear energy. Swadesh and I went to Bryan, Texas, to present his work.

Rep. Flores was respectful, showed genuine interest, and asked questions. He wanted to understand the timeline to act on climate change and asked Swadesh, "How many years have

Fig. 4. Me with husband, Swadesh Mahajan, and Representative Bill Flores of the Texas Congressional District TX-17 in his Bryan office.
Source: Photo from the author's collection

we got?" Swadesh explained that the time to act is now. Flores asked for a proposal for expanding on the invention, a proposal that was later sent to his office. Swadesh didn't hear back. Congressman Flores retired from Congress in 2020 after serving in the U.S. Congress for ten years. I believe that Rep. Flores asked a very important question: "How many years have we got?" I believe he knew the urgency of the crisis and wanted a scientist to confirm it.

CCL'S CARBON FEE AND DIVIDEND PROPOSAL

The Energy Innovation and Carbon Dividend Act (H.R. 5744) of 2023 has three major components per the Citizens' Climate Lobby (community.citizensclimate.org/resources/item/19/374).

1. Put a fee on fossil fuels like coal, oil, and gas, starting low at an initial fee of \$15/metric ton on the CO_2 equivalent emissions of fossil fuels, grow it every year

by $10 per metric ton, and impose it upstream—as near as feasible to the mine or oil well of extraction or port of entry. This will drive down carbon pollution because energy companies, leading industries, and American consumers will move toward cleaner, cheaper options.

2. The net revenue collected from the carbon is allocated in equal shares every month to the American people to spend as they see fit. Program administrative costs are paid from the fees collected. The government does not keep any money from the carbon fee.

3. To protect U.S. manufacturing and jobs, imported goods from countries with equivalent carbon policies will pay a carbon adjustment at the border, and goods exported from the United States to countries without similar policies will receive a refund.

The Climate Leadership Council,[8] made up of Republican leaders, is proposing a higher initial fee of $45/metric ton. Energy companies will transfer the higher costs to consumers. This debate needs to ensure that lower- and middle-class families can join others in the transitioning economy without being hurt.

About two-thirds of Americans will receive more in dividends than they will pay in higher prices (Fig. 2). This feature will inject billions into the economy, protect family budgets, free households to make independent choices about their energy usage, spur innovation, and build aggregate demand for low-carbon products at the consumer level. Climate pollution goes down as greenhouse gas emissions are reduced, improving the public's health and well-being. Savings in public health alone will be far more than the cost of the policy.[9]

This is just a starting point in turning the tide to lower carbon emissions. An important parallel step as part of the Carbon Dividend Act would be to plan new paths for workers who will be transitioning out of the fossil fuel industry. It would be a difficult transition and a challenging opportunity. Management of this transition should be done using training programs for workers who have powered the growth of the industry for decades and can now do that to usher in a clean-energy generation.

CARBON PRICING POLICIES AROUND THE GLOBE

According to the Citizens' Climate Lobby, as of May 2021, sixty-four carbon pricing policies were in operation and three scheduled for implementation in the world. These included both carbon taxes and emissions-trading schemes. These policies cover about twenty-two percent of worldwide emissions. The list of countries that already practice some method of national carbon pricing includes Argentina, Canada, Chile, China, Colombia, Denmark, the European Union (twenty-seven countries), Japan, Kazakhstan, Korea, Mexico, New Zealand, Norway, Singapore, South Africa, Sweden, the United Kingdom, and Ukraine. Other countries that are considering joining them include Brazil, Brunei, Indonesia, Pakistan, Russia, Serbia, Thailand, Turkey, and Vietnam. Many countries, including most of the larger trading partners of the United States, have instituted some form of national carbon pricing.

China, the country with the biggest carbon footprint as of 2021, has started eight regional greenhouse gas emissions trading projects since 2013. In July 2021, they launched a nationwide expansion within their electricity sector, with other sectors to be integrated over time.

Some of Canada's provinces have carbon taxes, starting

with British Columbia in 2008. It has now launched its federal backstop carbon-tax-and-dividend program for provinces that have not yet enacted carbon pricing, and this policy can reduce emissions to thirty percent below 2005 levels by 2030.

Of all the world's developed economies, only the United States and Australia do not have nationwide carbon pricing in place. Globally, if you look at the twenty biggest economies, India and a few other Persian Gulf countries are the only other countries besides the U.S. and Australia that do not have carbon pricing.

To build support for the federal policy, CCL volunteers (among their other to-dos) reach out to state and local governments, universities, businesses, institutions, and community groups. They set up tables at local events and advocate for policies that reduce greenhouse gas emissions and for community bridge-building—both with our elected officials and within our communities.

CITY OF AUSTIN CARBON FEE AND DIVIDEND RESOLUTION

As an early leader in adopting green building policies, the City of Austin had already approved climate-related resolutions before the CCL's Austin chapter volunteers approached the city council. We were able to get Carbon Fee and Dividend Resolution 20180426-37 passed by the City Council. This resolution showed Austin's support for the Carbon Fee and Dividend policy. As the city leadership expressed in writing, "A steadily increasing revenue-neutral carbon fee and dividend would be minimally disruptive to the economy while sending clear and predictable price signals to individuals and businesses purchasing and using carbon-based energy resources." As part of this resolution, the city leadership asked the U.S. Congress to adopt policies to reduce the impacts of climate change.

ADVOCACY IN THE STATE OF TEXAS

The State of Texas is a well-known leader in oil and gas production, but Texas also leads the nation in wind power generation and is almost at the top in the generation of solar power. Unfortunately, Texas is the largest emitter of greenhouse gases in our nation. Without Texas, our nation cannot decarbonize itself. If Texas were a country, it would be the eighth-largest economy[10] and the eighth-largest emitter of greenhouse gas emissions in the world.[11]

With these facts in mind, the CCL leadership decided to approach the Texas State Legislature in 2023, during the session they hold once every two years for 140 days for enacting laws. This means that the Texas Legislature meets for 140 days out of a total of 730 days unless the governor calls a special session. Constituents have difficulty bringing up issues to state leaders when all the work of committees and two state houses must sort through the number of bills filed in a very short time. Important issues remain unaddressed.

Policy decisions made by the Texas Legislature affect the energy industry as well as individual Texans. How we generate and transport energy has a profound impact on our environment and quality of life, which is why CCL's 10,000 Texas volunteers asked lawmakers to act on improving our environment. Methane leaks from oil and gas production in the Texas Permian Basin are a major concern. On Texas Lobby Day, March 28, 2023, eighty-one CCL volunteers had sixty-six meetings with state representatives, out of which forty-seven were with Republicans and nineteen were with Democrats. I was part of four meetings with representatives, including my State Representative for District 50, James Talarico.

Texas continues to produce record amounts of energy from solar and wind, but to transmit it to homes and businesses, the state needs improvements to the grid. The backlog still exists, and transmission lines are not fully fixed. The generated energy backlog and transmission bottlenecks such as these cost Texans over $5 billion in 2022 and 2023.[12] Consumers, the State, and the environment—all of them—will benefit if lawmakers enact the right policies.

CCL's Primary Asks of Lawmakers in 2023:

- Keep Texas's competitive edge as a leader in producing wind and solar power.

- Expand economic opportunities in Texas.

- Improve grid reliability.

- Save consumers money.

- Reduce greenhouse gas emissions.

- Expand bipartisan participation in the Texas Energy and Climate Caucus.

CCL supports climate solutions that grow economic opportunities and are market-based, cost-effective, efficient, and implemented in a socially equitable manner. The climate crisis has brought already existing inequities to the forefront, making the disparities worse for those who have fewer resources to live through any type of climate event, whether flood, drought, wildfire, sea level rise, or extreme heat or cold. When unable to cope without economic means, this crisis causes enormous physical and psychological stress, affecting public health and degrading social relationships.

Most American people are concerned about global warming.[13] For our own good, we must pursue with determination, empathy, and wisdom for all the young people we love. When an old man plants a tree today, he is doing it to benefit others in the future. How can you influence your community and nation and join the global efforts to address the crisis?

CHAPTER 9

WE CAN AND MUST ACT: IT'S UP TO US

PROTECTING PEOPLE AND OUR ENVIRONMENT

As professionals in the building industry, we provide services to either public or private sector owners and carefully plan the spending of funds to get the maximum benefit for owners. In the same way, for us to be caring guardians of the planet, we must spend public and private funds wisely on actions to address the root causes of the climate crisis. The scientific facts that I shared in the previous chapters show the impact of building activities and other sectors on the environment and people around the globe. When extreme weather events hit hard, we have very little choice at the time but to adapt. But to find long-term solutions to the climate crisis, we need mitigation strategies.

A large amount of carbon dioxide is already accumulated in the atmosphere, and we continue to add additional carbon dioxide, methane, and other greenhouse gas pollutants every year. A portion of that gets absorbed by the Earth's systems, like forests, landscapes, and oceans, known as carbon sinks, but that is not enough to stop an overall increase. According to scientists, we have passed the time to stabilize our climate by just reducing emissions to net zero. Pursuing net-zero emissions for all sectors, including buildings, is crucial. It is equally important to find ways to remove the accumulated atmospheric carbon to stabilize our climate.

Professionals use a critical path analysis method to prepare construction schedules. The same concept could be applied to address the climate crisis effectively and wisely by listening to experts who prepare models and make predictions for future weather.

How do governments and private investors join forces to prioritize investments in clean and safe energy in thoughtful and equitable ways? Public entities must take the lead and provide certainty and signal to the markets to invest in clean energy as well as innovations in the removal of carbon dioxide to cool the environment. Such efforts are already happening across the globe, but they must be expedited to address the urgency this crisis deserves.

Project teams design and construct green buildings and provide all necessary support for the operation and maintenance of buildings. More than half the emissions coming from the building sector are from after buildings are occupied. Therefore, occupants must understand what is needed to lower the emissions. Professional and personal actions combined would go a long way.

Your circle of influence is larger than you think!

Fig. 1. You have a critical part in influencing your community, city, state, and nation. All combined, those actions would have a major impact on the global community.

YOU AND YOUR ACTIONS

- Keep in mind that electricity and water are the most precious commodities we have, and we must conserve them.

- Weatherize and insulate your home to lower heat gain in summers and heat loss in winters.

- Combine your use of fans and natural ventilation in good weather to cut down on air-conditioning and heating.

- Save money by setting your thermostat a few degrees higher in the summer and a few degrees lower in the winter. Notice a difference on your next bill that comes with changing by even one degree.

- Change to LED lighting while replacing light bulbs. Replace your appliances and equipment with electric models when possible, using incentives in the Inflation Reduction Act of 2022. An electric stove is a major health benefit by stopping methane leaks from a gas stove.

- Plant native trees, native flowering plants, and wildflowers instead of covering your property with water-consuming grass.

- Buy local foods, and only as much as you will eat for the week. Try plant-based foods as much as possible.

- Tired of wasting food? Wash and freeze leftover fruits for smoothies and sorbets. Toss frozen fruits into oatmeal and pie filling—oh...and on top of ice cream. Freeze veggies like tomatoes, carrots, and spinach for easy one-pot meals, lasagnas, and casseroles.

- Compost to reduce landfill waste and methane pollution. In 2021, landfill waste was responsible for fourteen percent of methane emissions.[1] Composting reduces the need for more landfill space and the transportation costs to access landfills, which are far from where the waste is collected.

- Use gasoline-free transportation, including bikes, public transit, and electric vehicles.

- Turn to a more actionable items list prepared by the City of Austin's Office of Sustainability (Sustainability | AustinTexas.gov),[2]

- Reduction of waste, including plastic bags, wrappings, Styrofoam containers, non-reusable plates and cups,

Fig. 2. Environmental Protection Agency – Wasted Food Scale,[3] October 2023. Starting from dark shades on the left side to lighter shades are preferred options in terms of priority.

paper, aluminum cans, old clothes, cardboard, and food waste, is doable today. Composting can greatly reduce landfill waste for others, as it has done for me.

YOU AND YOUR COMMUNITY

Learning about scientific facts inspired me to act. I once started legal action with my homeowner's association (HOA) because I wanted to install solar panels and was not allowed to do that by the Architectural Control Committee. Incidentally, the State of Texas had taken legislative actions at that time, allowing homeowners to install solar panels. I became a member of the Architectural Control Committee and our committee prepared guidelines to install solar panels. Soon we had them installed with consultation with Austin Energy for weatherization and available rebates.

When you hear about a climate disaster in your area, seek out the facts from those who study them on a day-to-day basis, including news meteorologists. Find out how you can get involved or rally your community to action. Talk to people on walks in your neighborhood, in meetings, over coffee or dinner, before or after worship service, at work, and at social events. According to climate scientist Katharine Hayhoe, "The most important thing you can do to fight climate change: talk about it."

YOU AND YOUR CITY AND STATE

Reach out to your city leaders who can influence planning, zoning, and building code decisions to help select low-emissions options for construction and transportation. The City of Austin requires construction waste management in all its projects.

When we worked on sustainable guidelines, construction waste management was one of the first items. You could persuade your city to follow such sustainable practices. Some cities own electric utilities. You can, for example, persuade your utility company to cut down on fossil fuels and invest in clean energy. Bring attention to what you want to happen in your area, whether it's cutting down on fossil fuels, establishing better composting and recycling options, increasing green space, or reducing impervious ground surfaces. If you want a benefit where you live, start working with your local government.

Your state legislators are the people who routinely make decisions that affect your life. Call or write to your state representatives on a regular basis. Letters, phone calls, and in-person meetings are more effective than emails. Constituents' views matter to lawmakers. They pay attention to your concerns, especially if you are a professional and can speak to them about improvements to building codes or investments in renewable energy.

Action is easier and more effective when taken in concert with other environmental groups. Many environmental groups testify at state committees on environmental actions. Be one of those voices or part of a volunteer research team. CCL's Austin chapter was successful in getting support from the City Council for the Carbon Fee and Dividend policy and asked the U.S. Congress to act on it.

YOU AND YOUR NATION

Reach out to your two senators and ask them to enact laws related to clean air and clean water that would reduce harmful and polluting emissions and enhance sustainability. Work with the Citizens' Climate Lobby, Sierra Club, or another environmental group listed in Appendix 4.

Many environmental groups are already working with members of Congress to take federal action. If constituents continue to ask for action for a better world for their children and young people, members of Congress are more likely to act. Better yet, vote. Elect leaders who will listen. Inspire your friends and family to do the same.

GLOBAL ACTIONS BY NATIONS

Use the power of your vote. You can elect leaders to the U.S. House and Senate who are willing to collaborate with the European Union and China. They are investing a lot more than the United States in clean-energy sources and research on materials. If nations can share research to produce materials that can cut down emissions, the whole world benefits and it is easier for building professionals to decarbonize buildings.

The climate crisis is not a partisan issue. Disasters affect everyone, no matter what your political preference. Organizations like the Intergovernmental Panel on Climate Change (IPCC) can only be effective if nations follow through with their pledges.

At the Conference of Parties, when nations meet, they make pledges to reduce emissions. They have met twenty-eight times, but the progress that resulted from those meetings has not reflected the urgency of the climate crisis. The pledges made were nonbinding, and no avenue exists that ensures nations act accordingly. Global engagement is difficult but crucial, which is why your vote and voice are important.

Expressing your concerns at the ballot box remains the most important tool you have as a citizen in the United States. Your vote can help to ease the crisis for those who face the next catastrophic climate-related crisis, especially if it places them on the brink of losing their lives and livelihoods, which it often does. For those fortunate enough to have the means to live through climate disasters, the climate crisis still looms large in their lives. For example, melting glaciers take away the pleasure of visiting glacial parks for skiers and hikers. Melting glaciers affect the water supply, causing problems for people who are dependent on them. Air quality and pollution together create conditions that prevent people like me from spending time outside to enjoy nature.

The CCL has chapters in forty-six countries, and the environmental advocates in those countries are demanding policy changes.

Your circle of influence is larger than you think.
Your actions really do matter.

THE BIDEN ADMINISTRATION'S ACTIONS

President Biden described the climate crisis as an "existential crisis" in his inaugural speech. In 2021, President Biden and some of the congressional leaders were making a push for pricing carbon as part of a bill called Build Back Better, which failed to get enough votes. Consideration of carbon pricing itself was a result of the advocacy work of many environmental groups. Our group, CCL, held 920 meetings with congressional offices, generated 225,000 contacts to Congress, published over 2,000 letters to the editor, and published 676 op-eds in 2022.

President Biden had set an ambitious goal for the U.S. to reduce emissions to fifty percent by 2030.[4] In July 2022, there was a breakthrough, and Senate Democrats reached an agreement for a bill that included the largest investments ever made by the U.S. Congress to cut emissions and address the climate crisis. The bill was called the Inflation Reduction Act (IRA) of 2022, and it passed without a single Republican vote in the U.S.

House or Senate. People in the red and blue states benefit, and reports show that red states are taking advantage of the IRA.

Our advocacy can help both Republicans and Democrats join forces for the substantial climate action that our nation needs to be a global leader in solving this crisis. IRA was a remarkable bill with major investments to transition toward clean energy sources and other significant measures. The most important of all is the enhancements to Internal Revenue Service (IRS) 45Q, which greatly increased tax incentives for carbon capture. The policy now provides a direct cash payment instead of a tax incentive. If all states take advantage of funding in the IRA, we can reduce thirty-two to forty-two percent of emissions by 2030, according to some estimates.[5]

ACHIEVING HIGHEST BENEFITS FROM THE INFLATION REDUCTION ACT (IRA)

Besides signing the IRA, President Biden has signed two other major bills, the CHIPS and Science Act and the Infrastructure Investment and Jobs Act. They both contain measures to address climate change. Together, they are expected to make a major difference in lowering emissions by giving a boost in transitioning from fossil fuels to a clean economy and creating a major incentive to remove carbon dioxide from the sky. Lowering harmful emissions creates sustainability, and the removal of carbon dioxide creates climate restoration.

Here are the highlights of the IRA. Every action noted would create jobs:

- Clean electricity production and investment tax credits for wind, solar, and other renewables, especially a thirty-percent tax credit for residential solar

- Tax credits for new and used electric vehicles

- A federal fee on the potent greenhouse gas called methane (CH_4) to cut pollution and reduce global warming

- Incentives for oil and gas companies to fix methane leaks

- Natural climate solutions like forestry grants, wildfire prevention, forest conservation, forest management, grants for cities to grow urban forests, wood innovation grants, climate-smart agriculture, and conservation programs to store carbon in soil and trees

- Consumer home energy rebate programs like weatherization, heat pumps, and other electrical appliances to improve energy efficiency in homes, with higher rebates to low-income households

- Electric vehicles for the United States Post Office

- National green bank for green tech accelerator

- Investments in disadvantaged communities to address environmental justice and equity issues

- Domestic clean-energy manufacturing and granting the authority to the Department of Energy to lend funds to new green companies and technologies

- Tax credit for carbon sequestration, production tax credit for nuclear power, extended tax credits for biodiesel fuels, and tax credit for aviation biofuel—all to reduce emissions and for low-CO_2 hydrogen production

- Permanent coal tax extension to fund the Black Lung Disability Trust Fund

- And, most important of all, a no-pay-limits carbon dioxide removal incentive to stabilize our climate

U.S. environmental groups are persuading their lawmakers to support the IRA and the permitting reforms necessary to take maximum advantage of the IRA's incentives to accelerate the transition to clean energy.

FOSSIL FUELS AND ENERGY NEEDS

We have all enjoyed the benefits of modernization powered by fossil fuels in countries with advanced technological development. In developing countries like India, rural communities are beginning to enjoy the benefits of electricity. Many in the United States, including myself, have cushy lives and have acquired many material possessions in comparison to our parents and grandparents. With the growth in manufacturing and rise in income levels for the affluent, our homes have become larger. I admit to having a large house that was purchased in 1999, before my awareness of how our high-consumption lifestyles affect the environment. Doing volunteer work with the Citizens' Climate Lobby has opened my eyes to the effects of our activities on the environment. We have incorporated all energy-saving features, including solar panels on the rooftop, to keep net electricity use to a minimum. Warming has an impact. Fourteen years ago, solar panels met our electrical needs almost completely. But now, with an extended summer season with record-breaking temperatures, we have been using a higher amount of electricity. Electric cars have helped to cut down on gasoline. Being a vegetarian is not difficult with my Jain upbringing, but we have reduced packaged foods and now eat as many fresh foods as possible. I am fully cognizant of the fact that our personal actions are not enough. But they are a good starting point for oneself to act as shared in the model above and to inspire others.

According to the Energy Institute (https://www.energyinst. org/statistical-review), the world's energy consumption and emissions rose in 2023. We admittedly need a lot of energy specifically for peak-hour demand, and so far, it largely comes from coal, oil, and gas. Solar and wind-power generation are making impressive gains, but we have a long way to go to satisfy our 24/7 needs and vastly increasing demands from electrification amid the rapidly growing electric vehicle (EV) market and increase in data centers. We have an urgency to find other clean energy sources to decarbonize and electrify all sectors of the economy, including buildings. By implementing all the green building strategies discussed earlier, we can reduce the amount of electricity needed, leading the utility company to lower peak load and emissions in the building sector. Electrification will increase electricity demand above and beyond efficiency gains with sustainability practices.

According to the online U.K. newspaper *The Telegraph* (April 7, 2023), China is investing in huge solar farms and other nonpolluting energy sources. China added 217 gigawatts of solar capacity in 2023, a figure that surpasses its previous record of eighty-seven gigawatts in 2022 and exceeds the "entire solar fleet of 175 gigawatts in the U.S."[6] To make sense of these numbers, let's compare each gigawatt to a typical nuclear power plant, which generates one gigawatt of electricity. It is like China adding the equivalent of 217 nuclear power plants! Now, investors in the United States, Europe, India, and other countries are accelerating investments in clean energy to compete with China.

U.S. military bases are impacted by extreme weather, and they are preparing studies and heavily investing in solutions to address the military's operations. The U.S. military has bases around the world, and they have seen the effects of climate change on sea level rise, extreme heat, climate refugee migrations, wildfires, and flooding.

Research is ongoing to make fusion power a reality, but it is not feasible yet. There are possibilities to repurpose coal power plants by converting them to biomass fuel, switching to renewable power, or adding small modular reactors.[7] I strongly believe that training programs for workers in the fossil fuel industry must be planned in the clean energy transition. Pollution is affecting the health of coal miners and workers in the fracking industry. Allergies and asthma caused by pollution are increasing among children. We need sustainable practices for the health of everyone.

Scientists are predicting a much hotter world. That is going to affect all of us and our future generations. My childhood stories speak about the difficulties and stresses of dealing with heat, and that would be even more difficult as the global temperature curve continues to rise. Looking back, we had a much more stable climate with predictable seasons. Who doesn't want a stable climate? Most Americans are concerned about global warming and an even larger percentage of people in India are worried.[8]

YOUNG PEOPLE'S MOVEMENT

CCL encourages young people to form groups to support policies to address climate change. It is truly inspiring to see young people's organizations in the U.S. In my lobby meeting on June 11, 2024, it was truly inspiring for me to work with a seventeen-year-old high school student from Lubbock, Texas, who felt he had to act after he witnessed the devastation caused by Winter Storm Uri. There is an organization called High Schoolers for Carbon Dividends.[9] It was co-founded by 700-plus leaders from all fifty states, including national winners of debates, science and economics competitions, Scripps National Spelling Bee champions, student government presidents, and many more. Also, the student body presidents at more than 450 colleges and universities support this policy and have issued a bipartisan statement in support of a Carbon Fee and Dividend policy.

As young people with decades of life ahead of them, they are clear-eyed about what climate disruption means for their generation. According to Students for Carbon Dividend (S4CD),[10] "That's why we recognize the power of a consensus solution like carbon dividends to bridge partisan divides, protect our shared environment, and strengthen the economy. Where our political leaders have been unwilling, or unable, to forge agreement around common-sense solutions, we on college campuses are showing them how it's done."

Greta Thunberg, a Swedish environmental activist, started her sit-outs, "Fridays for Future." She has inspired millions of students and has raised awareness among young people about the urgency of the climate crisis. She has participated in global conferences and has courageously confronted powerful decision-makers in governments and big corporations.

THE DISCONNECT MUST STOP

This disconnect between what people want their leaders to do and what the leaders end up doing was demonstrated in the 2023 legislative session of Texas. Big oil, coal, and gas companies continue to make campaign contributions to politicians in Texas and other states. These contributions influence lawmakers of both parties in the House and Senate to provide subsidies to build new gas power plants and slow down the generation of solar and wind power by putting more regulations on clean energy-generating sources, like solar and wind.

It is mind-boggling that our elected leaders disregard the fact that solar and wind power have generated billions of dollars for Texas, have reduced electricity costs for Texans, and have helped sustain rural communities and school districts around the state. Some of those lawmakers are ignoring another obvious benefit of solar and wind power: the reduction in harmful emissions and improvements to the health and well-being of Texans. It was distressing to see the 2023 Texas Legislature's attempt to slow down Texas's burgeoning solar and wind generation industries.

WHERE DO WE GO FROM HERE?

Here is an appeal for a climate that is stable and that would allow us—you, me, and millions of young and vulnerable people—to enjoy the environment, clean air, and clean water. These are policies we can get enacted:

- Removal of carbon and other greenhouse gas emissions from buildings by enhancing sustainability in everything in planning, design, and construction and after occupancy

- Practicing sustainability in all our activities

- A path to electrification of everything, which would increase electricity consumption, requiring additional generation capacity by renewable and other clean energy resources

- A path of removal of gigatons of already existing pollutants in the environment to cool it

- Reaching out to lawmakers to bring their attention to the worsening climate crisis

- Protection of communities suffering from the effects of global warming

- Engagement of communities and people as part of all solutions

My grandchildren, Kimaya and Kavi, the grandchildren of my three brothers, and the other young people I love may not be able to enjoy clean air, clean water, and the beauty of the natural environment that our generations have enjoyed. They may be victimized by extreme weather events and not realize fulfillment in life due to our indulgences and inability to take collaborative actions. We can and must make every effort to lower the pollutants and emissions and restore our climate. As a stanza from Swadesh Mahajan's poem, "Mother Earth Trilogy" (see Appendix 5), goes, "It took a while, but the idea spread earth is to be exploited to enrich the man his needs, wastefulness, luxuries unbound greed and arrogance the dominant theme unlimited perpetual growth rose in ranks of divinities that fully control the clan."

I have started taking actions on the Circle of Influence model (Fig. 1 in this chapter) that I created. I do my actions in every circle and encourage my colleagues in the building industry and environmental groups to join me. These actions are meant for the people in the United States, but they can be applicable anywhere. Depending on where you live, you may have limitations or many other opportunities. My hope is that you will start your own list using this model.

We know that emissions resulting from transportation are very high in the U.S. Like other cities in Texas, Austin is spread out and without the convenience of a well-connected and effective public transit system that many would like if it were available. Austin has been trying to get voter approval for a revamped public transit system, specifically one that includes a comprehensive light rail system, since 2000. Finally, in 2023, we are seeing some progress. Preliminary plans for light rail are progressing along with funding, but, based on my experience in the construction of public facilities, it will be some time before Austinites will see a well-functioning light rail. As of this writing, the progress has hurdles due to court action. The longer we wait, the more expensive and difficult it will be for residents. Imagine having a public transit system in our city had it been approved twenty-four years ago! That dream would have alleviated traffic congestion and improved air quality and the quality of life in general. Those of us who supported it then continue to push for it now. When we elect leaders who care for the health and well-being of their constituents, they will support policies to lower pollutants and improve the air we breathe and the water we drink.

When we combine the innovative talents of so many professionals, we have a huge potential to find ways to plan, design, and construct buildings and infrastructure using new innovative techniques and materials. Once we focus on policies to do just that, the possibilities are immense.

One thing is clear—we cannot continue the path of business as usual.

The building industry involves a very large group of people in public and private sector projects. There are many who have done outstanding work, like Edward Mazria of the AIA's 2030 Challenge does by keeping the climate crisis front and center of his practice and for the AIA. Those who are still waiting to join the "Green Buildings and the Climate Crisis" movement, I sincerely hope that they will see the connection between our work and the looming climate crisis. It is magical to create something that would fulfill our obligation but also restore our climate and enhance the quality of life for us and future generations. Most importantly, we can all continue to talk to friends and family about our changing environment.

Let us be a part of creating a livable world for all.
It is up to us to build green buildings and a political will to restore our climate.
Let's start actions at home and then let them travel to other parts of the world.

ACKNOWLEDGMENTS

My grandparents inculcated the value of conservation in me. I am indebted to them.

My parents, Ramesh P. Sutaria and Veena R. Sutaria, for their unconditional love and support, which has been and still is a guiding force after they are long gone.

Swadesh Mahajan, my husband, who has always been there by my side, inspiring and loving. It has always been illuminating to discuss any topic with him, not just scientific, to enhance understanding of life in general. He has enriched my life with his sharp intellect, sense of humor, and concern for our environment. He has been a physicist and a professor all his life.

Thanks to my stepsons, Romi Mahajan and Rahul Mahajan, whose critical comments in different stages of the manuscript were most encouraging and helpful. Romi and his better half Parul Shah are particularly concerned about the effects of global warming on their children—our grandchildren—Kimaya and Kavi. Thanks to family members Sharmila Roy, Zia Hasan, Rafeeq Hasan, and Christine Welcher for moral support.

Thanks to the Sutaria family in India and in the U.S. and the Mahajan family for their warmth and encouragement. Special thanks to my brothers, Prashant, Shrikant, and Jagdip Sutaria, who routinely shared weather news and events and their effects on the lives of people in India. They are already suffering from deteriorating air quality in Ahmedabad today and find it hard to imagine the kind of environment their grandchildren will live in when global warming further worsens the air quality. They have been distressed by the cutting down of mature trees to make way for luxury apartments, bridges, or roads in Ahmedabad, but they feel helpless. Thanks to my aunt Suhas Shah and her family, who shared stories of the old city homes in Ahmedabad with me.

Architects Madhavi Desai and Miki Desai in India have been my friends since the School of Architecture days in Ahmedabad. We were students at the University of Texas at Austin at the same time. Their comments and guidance were invaluable. I can't thank them enough.

I am deeply grateful to my friend Karen J. Wilson, who read different versions of the manuscript and provided her insights throughout this journey. My friends in Austin kept me entertained, especially Veena Gondhalekar, Suman Olivelle, and Patrick Olivelle. Patrick is an outstanding historian who offered pointers for first-time writers like me.

Friends in the architecture and engineering profession—Toni Thomasson, Patti Jordan, Gail Vittori, Riley Triggs, Kristine Walker, Adele Houghton, and Sarah Talkington—provided thoughtful comments after reading all or parts of the manuscript. For that, I am truly grateful.

Thanks to all my friends at the Ahmedabad School of Architecture who boosted my spirits, especially the group of female architects. They all continue to make contributions to sustainable architecture.

I am thankful to Bruce Melton, Anna Graybeal, Hamilton Richards, Susan Adams, Marie Miglin, Carolyn Appleton, and many others of the Austin chapter of the Citizens Climate Lobby who read the manuscript at some stage and gave valuable feedback. Special thanks to Bruce Melton, whose passion for his work and deep understanding of the climate crisis were refreshing and opened my mind to a wide range of solutions.

Thanks to my colleagues at the city, the ones doing their best to advance sustainability and others with whom I have enjoyed working over the years. It was a learning experience and an opportunity to have thought-provoking discussions. Special thanks to the previous city manager of Austin, Toby Futrell, who listened to us and supported our efforts for adoption of LEED and sustainable building guidelines. It was a pleasure to work for her.

Austin Energy Green Building (AEGB) Program has been committed for many years and continues to serve the Austin community by disseminating information and organizing workshops and seminars to find solutions that advance sustainability and moving toward a renewable energy future. I have enjoyed working with AEGB professionals over many years, and I am thankful to all of them.

Thanks to design consultants and contractors who worked with me on the city's public facility projects and made their best efforts to implement sustainability despite concerns for project budget.

I am most thankful to climate scientists Dr. James Hansen and Maki Takeichi, who were most generous in providing scientific data and graphs that were critical in making my case to fellow professionals and climate advocates.

NOTES:

PREFACE

1. Rohde, Robert (2024), Global Temperature Report of 2023, https://berkeleyearth.org/global-temperature-report-for-2023/

2. International Energy Institute, https://www.energyinst.org/statistical-review

3. Yale Program on Climate Change Communications, https://climate communication.yale.edu/publications/climate-change-in-the-american-mind-beliefs-attitudes-fall-2023

4. Intergovernmental Panel on Climate Change, https://www.ipcc.ch/assessment-report/ar6/

5. Barber, Daniel A., *Modern Architecture and Climate, Design before Air Conditioning*, Princeton University Press, Princeton, and Oxford.

6. United Nations, Academic Impact, https://www.un.org/en/academic-impact/sustainability

7. Austin Energy Green Building Program, https://austinenergy.com/energy-efficiency/green-building

8. Morton, B. (2011), "Falser Words Were Never Spoken," *The New York Times*, opinion page contributor.

CHAPTER 1: EXTREME WEATHER EVENTS

1. National Oceanic and Atmospheric Association, https://www.ncei.noaa.gov/access/billions/

2. Berkeley Earth, https://berkeleyearth.org/global-temperature-report-for-2023/

3. National Oceanic and Atmospheric Association, https://www.noaa.gov/news-release/warmest-arctic-summer-on-record-is-evidence-of-accelerating-climate-change

4. Intergovernmental Panel on Climate Change, https://www.ipcc.ch/sr15/

5. Climate Central, https://www.climatecentral.org/climate-matters/urban-heat-islands-2023?

6. https://www.weather-atlas.com/en/india/ahmedabad-climate

7. Dr. Stephen J. Pyne, Professor at the Arizona State University, specializing in the history of fire, https://en.wikipedia.org/wiki/Stephen_ J._Pyne

8. National Oceanic and Atmospheric Association, https://www.noaa.gov/climate

9. National Oceanic and Atmospheric Association, https://marinedebris.noaa.gov/emergency-response/hurricanes-harvey-irma-and-maria

10. Collier, Chelsea, "Get Smart," *Texas Architect*, May/June 2023.

11. Article in *The New York Times* on September 29, 2022, "How Hurricane Ian became so powerful," by Elena Shao, Nadja Popovich, and Mira Rojanasakul, https://www.nytimes.com/interactive/2022/09/29/climate/hurricane-ian-florida-intensity.html

12. Dr. Steven Clemens, Department of Earth, Environment and Planetary Sciences, Brown University, https://vivo.brown.edu/display/sclemens# Publications

13. Sheerazi, Hadia A., State of the Planet, "The Flood Seen From Space: Pakistan's Apocalyptic Crisis," September 12, 2022, https://news.climate.columbia.edu/2022/09/12/the-flood-seen-from-space-pakistans-apocalyptic-crisis/

14. https://www.cnn.com/2023/04/12/weather/florida-flash-flood-fort-lauderdale/index.html

15. National Oceanic and Atmospheric Association, https://www.noaa.gov/noaa-wildfire

16. Scripps Institute of Oceanography, https://scripps.ucsd.edu/news/broken-record-atmospheric-carbon-dioxide-levels-jump-again

17. Ozone (O_3) - https://en.wikipedia.org/wiki/Ozone

18. *Austin American* Statesman article, October 15, 2022, "City sees record 25 Ozone Action Days."

19. Climate Central, https://www.climatecentral.org/climate-matters/ozone-pollution-the-good-the-bad-and-the-dirty

20. IQAir, https://www.iqair.com/us/world-air-quality-report

21. International Center for Integrated Mountain Development in Kathmandu, https://www.icimod.org/

22. David Finlay Breashears, https://en.wikipedia.org/wiki/David_Breashears

23. https://science.nasa.gov/science-research/earth-science/greenlands-thinning-ice/

24. *KVUE News*, Austin, https://www.kvue.com/article/news/local/winter-storm-uri-report-deaths-texas-austin/269-bb023e53-5b35-492c-9baf-270c32a79cbd

25. *BuzzFeed News*, https://www.buzzfeednews.com/article/peteraldhous/texas-winter-storm-power-outage-death-toll

26. The American Society of Civil Engineers, https://www.texasce.org/wp-content/uploads/2022/02/Reliability-Resilience-in-the-Balance-REPORT.pdf

27. https://www.climate.gov/news-features/blogs/beyond-data/2023-historic-year-us-billion-dollar-weather-and-climate-disasters

28. https://www.texastribune.org/2023/11/30/texas-homeowner-insurance-climate-change-costs/

29. *National Public Radio News*, Oct. 6, 2022, https://www.npr.
org/2022/10/ 05/1126900340/florida-community-designed-
weather-hurricane-ian-babcock-ranch-solar

CHAPTER 2: BUILDERS OF THE PAST SHOWING THE WAY FOR THE FUTURE

1.   UNESCO, Word Heritage Sites, https://whc.unesco.org/en/
list/1645

2.   Joseph, Tony (2021), *Early Indians—The Story of Our Ancestors and
Where We Came From*, Page 123, Jugernaut.

3.   Joseph, Tony (2021), *Early Indians—The Story of Our Ancestors and
Where We Came From*, Page 124, Jugernaut.

4.   Joseph, Tony (2021), *Early Indians—The Story of Our Ancestors and
Where We Came From*, Page 194, Jugernaut.

5.   Vernacular Architecture, https://en.wikipedia.org/wiki/
Vernacular_architecture, https://en.wikipedia.org/wiki/Bernard_
Rudofsky, https://artincontext.org/vernacular-architecture/

6.   Laurie Baker, "Architectural Maverick: The Legacy of Laurie
Baker," https://www.re-thinkingthefuture.com/know-your-
architects/a10278-architectural-maverick-the-legacy-of-laurie-
baker/

7.   Desai, Miki, (2019), *Wooden Architecture of Kerala*, Mapin Publishing
Gp Pty Ltd.

CHAPTER 3: CHILDHOOD HOME IN INDIA

1.   Jain practices described in this book are what I learned from my
grandfather, elders in the family, and the Jain *Sadhus (religious
leaders)* who routinely visited our house to collect food as part of

the Jain practices. I do not speak for the official Jain religion as described in the Jain scriptures.

2. https://www.cnn.com/travel/article/ahmedabad-first-unesco-city-india/index.html

3. Nighoskar, Devyani (2017), "Amdavadi pols: The doors that still welcome you," https://www.makeheritagefun.com/pols-of-ahmedabad/

4. Hawken, Paul, editor (2017), *Drawdown: The Most Comprehensive Plan Ever Proposed to Reverse Global Warming*, overall ranking 21 out of 80 items, "Clean Cook Stoves," pages 44 & 45.

5. https://assets.publishing.service.gov.uk/media/57a08b32ed915d3cfd000bc8/Status_Report_on_Use_of_Fuelwood_in_India.pdf

6. https://www.ncbi.nlm.nih.gov/pmc/articles/PMC7312255/

7. https://www.usda.gov/media/blog/2022/01/24/food-waste-and-its-links-greenhouse-gases-and-climate-change

8. Hawken, Paul, editor (2017), *Drawdown: The Most Comprehensive Plan Ever Proposed to Reverse Global Warming*, overall ranking 3 out of 80 items, "Reduce Food Waste," pages 42 & 43.

9. United Nations Environment Program, https://www.unep.org/news-and-stories/story/what-you-need-know-about-plastic-pollution-resolution

10. https://letstalkscience.ca/educational-resources/stem-explained/polystyrene-pros-cons-chemistry

11. Bloomberg, https://www.bloomberg.com/news/features/2023-09- 29/us-store-drop-off-plastic-recycling-often-ends-up-in-landfills? cmpid=BBD093023_GREENDAILY&utm_medium=email&utm_source=newsletter&utm_term=230930&utm_campaign=greendaily& embedded-checkout=true

12. Hawken, Paul, editor (2017), *Drawdown: The Most Comprehensive Plan Ever Proposed to Reverse Global Warming*, overall ranking 6 out of 80 items, "Women and Girls—educating girls," pages 80 to 82.

13. Gandhi Ashram in Ahmedabad, https://www.gandhi ashramsabarmati.org/en/

14. The Quit India Movement: This movement was a civil disobedience movement in India launched in August of 1942 in response to Mahatma Gandhi's call for the immediate independence of India from the British government. It was passed at the Bombay session of the All-India Congress Committee.

15. Khadi is a hand-spun cotton natural fiber that was promoted by Mahatma Gandhi during the freedom struggle of the Indian subcontinent. He started spinning the "*charkha*," or the spinning device, at the Sabarmati Ashram in Ahmedabad in 1917-1918. Mahatma Gandhi's image of spinning was omnipresent when he was photographed throughout the freedom struggle, which lasted until 1947.

16. Sari, There are different types of saris and descriptions. This is my description of saris worn by all elderly women in my house. It is a five-yard piece of cloth and can be made from different types of fabric, including cotton and silk. It is not stitched for any one person, and can be worn by anyone. Most of the time, my mother and grandmother wore a thin cotton sari to deal with heat. Women wear a blouse stitched for each individual woman to match the sari. A petticoat, like a long skirt, is worn to tie the sari on the waist. I have seen women who could walk very fast wearing a sari since the petticoat was wide enough to allow free movement. My parents never asked me to wear a sari until I was in college, and that was only on special occasions like weddings. Special occasion saris are generally silk with colors and designs. Even when the saris wear out, they are either given away to people in need or the cloth is put to some use.

CHAPTER 4: STEPPING INTO THE WORLD OF ARCHITECTURE

1. Balkrishna V. Doshi School of Architecture in Ahmedabad was designed by the world-famous architect Balkrishna Doshi, who played a major role in founding the school and became its first director. Doshi and other faculty members were trained in the U.S. and Europe, and the curriculum is influenced by their training in the U.S. Doshi devoted his life to the advancement of architectural education and the construction of many outstanding projects. Balkrishna Doshi was a recipient of the Pritzker Prize in 2021.

2. https://www.re-thinkingthefuture.com/rtf-architectural-reviews/a3347-theory-in-architecture-form-follows-function/

3. Barber, Daniel A. (2020), *Modern Architecture and Climate, Design before Air Conditioning*, "Architecture, Media and Climate," Princeton University Press.

CHAPTER 5: SUSTAINABILITY AND GREEN BUILDINGS

1. https://www.energystar.gov/about/how-energy-star-works/history

2. Austin Energy Green Building Certification, https://austinenergy.com/energy-efficiency/green-building

3. United States Green Building Counci, usgbc.org

4. Leadership in Energy Environmental Design, https://www.usgbc.org/leed

5. Office of Sustainability, https://www.austintexas.gov/department/sustainability

6. Capps, Kriston, "Biden Rolls out Minimum Energy Standards for Affordable Housing," April 25, 2024, Bloomberg News.

7. Major project team contributors: HDR Engineering Inc. was the prime consultant with sub-consultants listed here: Architectural - Kinney and Associates + Carter Design Associates, Lighting - Archillume Lighting Design, Inc., Hicks and Company, Structural Engineering - Jose I. Guerra Inc., Landscape Design - Winterowd and Associates, Permitting - Site Specifics, Geotech - Fugro South Inc., and TETCO. General Contractor: Jay-Reese Contractors, Inc.

8. Major project team contributors: Prime Architectural Consultant - Carter Design Associates, Associate Architects - Kinney and Associates, Associate Architects for the Library - The Lawrence Group Architects, Mechanical, Electrical and Plumbing Engineers - Tom Green and Company Engineers Inc., Structural Engineers - Jaster-Quintannia and Associates, Civil Engineers - Raymond Chan and Associates, Interior Design - Laurie Smith Design Associates, Landscape Design - Eleanor McKinney Landscape Architect, Inc., LEED Consultant - Sustainable Solutions, Lighting - Archillume Lighting Design, Inc., Acoustical Consultant - Dickensheets Associates, Permitting - Austin Permit Services, Cost Consulting - Hanscomb Inc., Community Outreach - ADISA Public Relations, Inc. General Contractor - Cadence McShane Corporation.

9. Major project team contributors: Jackson and Ryan Architects was a prime consultant with a team of sub-consultants listed here: Associate Architect - Parshall and Associates Architects, Mechanical, Electrical and Plumbing Engineers - Design Learned Inc., LEED Consultant - Monarch Design/Consulting, Structural Engineers - Jaster-Quintannia and Associates, Inc., Civil Engineering - Davcar Engineering, Landscape Design - Garcia Design Inc., Owner's Commissioning Agent - ACR Engineering, and General Contractor: VCC, LLC.

10. Major project team contributors: Prime Architectural Consultants - Lake Flato Architects, Library Consultants - Shepley Bulfinch Richardson & Abbott, Incorporated, LEED

and Ecosystem Consultants - HOLOS Collaborative, Building Enclosure - Simpson Gumpertz & Heger, Structural Engineers - Datum Engineers, Inc. and Datum Gojer Engineers, LLC, Other Structural Engineers - P.E. Structural Consultants, Inc., Reed Fire Protection LLC, Mechanical, Electrical and Plumbing Engineers - Encotech Engineering Consultants, Inc., Civil Engineers - Urban Design Group, Acoustical Consultant - Dickensheets Associates, Graphic Design - fd2s. General Contractor: Hensel Phelps.

11. The Living Building Challenge, www.living-future.org/LBC4.0

12. Josey Pavilion, Decatur, Texas, Lake Flato Architects, *Nature, Place, Craft, Restraint*, Page 72. The Dixon Water Foundation is the owner of the facility.

13. International WELL Building Institute, https://www. wellcertified.com/

14. Growing Global Waste, https://www.worldbank.org/en/news/ press-release/2018/09/20/global-waste-to-grow-by-70-percent-by-2050-unless-urgent-action-is-taken-world-bank-report

CHAPTER 6: BUILDINGS' IMPACT ON CLIMATE CHANGE

1. Environmental Protection Agency, epa.gov/ghgemissions/ overview-greenhouse-gases

2. https://www.climate.gov/news-features/understanding-climate/ climate-change-atmospheric-carbon-dioxide

3. Global Monitoring Laboratory, https://www.gml.noaa.gov/

4. Givoni, B. (1969), *Man, Climate and Architecture*, 2nd Edition, Page 90, "The Index of Thermal Stress."

5. Edison Electric Illuminating Company, https://www.nps.gov/ edis/learn/kidsyouth/the-electric-light-system-phonograph-motion-pictures.htm

6. https://www.modernfanoutlet.com/blog/the-history-of-ceiling-fans.html

7. John Gorrie,https://worldhistory.us/american-history/dr-john-gorrie-the-florida-ice-man.php#google_vignette

8. Carrier, Willis, https://www.britannica.com/biography/Willis-Carrier

9. Missouri State Building, https://www.ashrae.org/about/mission-and-vision/ashrae-industry-history/air-conditioning-and-refrigeration-timeline

10. American Society of Heating, Refrigerating, and Air-Conditioning Engineers, https://www.ashrae.org/about/mission-and-vision/ashrae-industry-history/air-conditioning-and-refrigeration-timeline

11. Hawken, Paul, editor (2017), *Drawdown: The Most Comprehensive Plan Ever Proposed to Reverse Global Warming*, overall ranking 1 out of 80 items, "Refrigerant Management," pages 164-165.

12. Montreal Protocol, https://www.unep.org/ozonaction/who-we-are/about-montreal-protocol

13. Kigali Amendment to the Montreal Protocol, https://www.unep.org/ozonaction/who-we-are/about-montreal-protocol

14. Climate Central, https://www.climatecentral.org/climate-matters/ozone-pollution-the-good-the-bad-and-the-dirty

15. Treehugger.com

16. Robles, Pablo, Holder, Josh, and White, Jeremy, (2023) *The New York Times* article, "How to Cool Down a City? Singapore is rethinking its sweltering urban areas to dampen the effects of climate change. Can it be a model?"

17. Residential Energy Consumption Survey (RECS), https://www.eia.gov/consumption/residential/

18. Nandi, Jayshree, October 2023, *Hindustan Times* article, "A/C use rising sharply in India, will surpass all other appliances by 2050: Report."

19. Arrhenius, Svante, (1896), https://theclimatehistorian.substack.com/p/svante-arrhenius-quantifying-the

20. https://www.gml.noaa.gov/ccgg/

21. https://scripps.ucsd.edu/

22. https://www.thomasedison.org/edison-quotes#:~:text=%E2%80%9CUnfortunately%2C%20there%20seems%20to%20be%20far%20more%20opportunity,and%20coal%20run%20out%20before%20we%20tackle%20that.

23. Sengupta, Somini, *The New York Times* article, February 28, 2023, "Climate Forward: The American Exception."

24. Neilson-Gammon, John (2021 and 2023 Reports), "Assessment of Historic and Future Trends of Extreme Weather in Texas, 1900-2036." https://climatexas.tamu.edu/products/texas-extreme-weather-report/index.html

25. https://www.cdc.gov/disasters/extremeheat/heattips.html

CHAPTER 7: REMOVING CARBON FROM THE PROCESS OF BUILDING

1. Major project team contributors: Architects and interior designers: Nimish Patel and Parul Zaveri of Abhikram, Environmental Consultants: Brian Ford, Short + Ford Associates, London, U.K. (Worked on typical laboratory block) and CL Gupta, Solar Agni International, Pondicherry, India (Remaining Blocks), Structural Consultant: Yogesh Vani Consulting Engineers, Ahmedabad, India, Utility Consultants: Dastur Consultant Pvt. Ltd., Delhi, India, Landscape Consultants: Kishore Pradhan, Mumbai, India, Lighting Consultants: Paresh Shah, Sukriti Design Incorporated, USA, Civil Contractors: Laxmanbhai Construction (India) Pvt Ltd., MB Brothers Ltd., Shetusha Engineers, and Contractors Pvt. Ltd, Materials Corner, JK Builders.

2. Chatham House, https://www.chathamhouse.org/2018/06/why-cement-major-contributor-climate-change

3. Portland Cement Association, https://www.cement.org/cement-concrete/ how-cement-is-made

4. Major project team contributors: Design-Builder Team of SpawGlass Contractors, Architects - MarmonMok Architecture, Civil engineering - Garza EMC, Mechanical, Electrical and Plumbing - Tom Green & Co. Engineers, Structural Engineering - Datum Rios, Environmental Consulting - Studio D Consulting + Design, Landscape - Asakura Robinson, IT and Security - Comb Consulting.

5. Agora Energy Transition, Success Stories, Switzerland Building Industry, Recycled Concrete and Low-Carbon Cement, https:// www. stadt-zuerich.ch/portal/en/index/portraet_der_stadt_ zuerich/environmental-strategy.html

6. Limestone Calcined Clay Cement, https://rmi.org/unleashing-the-potential-of-limestone-calcined-clay-cement/

7. Blue Planet, https://www.blueplanetsystems.com/

8. https://www.iea.org/reports/steel-and-aluminium

9. Bryson, Bill, (2010), *At Home: A Short History of Private Life*, Page 31.

10. https://www.statista.com/statistics/1154296/glass-production-eu/

11. https://ceramics.onlinelibrary.wiley.com/doi/full/10.1111/ijag.16674

12. RMI.org/buildings

13. https://brandassets.principal.com/m/b99311aee88e26/original/ Sustainability-Case-Study-Indeed-Tower.pdf

14. Wood Products Council, http://www.woodworks.org

15. sustainablecitycode.org

16. Edward Mazria, https://www.architecture2030.org/edward-mazria/

17. https://www.aia.org/design-excellence/climate-action/zero-carbon/ 2030-commitment

18. China Accord, https://www.architecture2030.org/international-and-chinese-firms-sign-historic-accord-to-tackle-climate-change/

19. https://regeneration.org/nexus/carbon-architecture

20. Hong Kong's Bauhinia Bud Building, https://www.worldarchitecture news.com/article/1726342/hong-kongs-leed-platinum-design-zaha-hadid-architects

21. https://savings.austinenergy.com/residential/offerings/home-improvements/weatherization

22. https://www.austintexas.gov/department/reclaimed-water-system

23. Lake Flato Architects—See Chapter 5 Notes, Note #10.

24. Net-Zero Buildings, https://www.buildings.com/resiliency-sustainability/decarbonization-net-zero/article/33037866/what-is-a-net-zero-building

CHAPTER 8: WHY IS ADVOCACY IMPORTANT?

1. Yale Program on Climate Change Communication, https://climate communication.yale.edu/publications/climate-change-in-the-american-mind-beliefs-attitudes-fall-

2. Paris Climate Agreement, https://unfccc.int/process-and-meetings/the-paris-agreement

3. Dr. James Hansen, https://www.columbia.edu/~jehl/

4. Carbon Fee and Dividend Bill, which is now called the Energy Innovation and Carbon Dividend Act, https://citizensclimate lobby.org/price-on-carbon/

5. https://en.wikipedia.org/wiki/Economists%27_Statement_on_Carbon_Dividends

6. https://citizensclimatelobby.org/about-ccl/citizens-climate-lobbys-founder/

7. *Nature* Article, "The hybrid returns": 25-28 News Feature Fusion MH CNS.indd.

8. Climate Leadership Council, https://clcouncil.org/

9. https://www.ncbi.nlm.nih.gov/pmc/articles/PMC6964422/

10. Texas Economic Development Corporation, https://business
    intexas.com/why-texas/economic-strength/

11. Environment Texas, https://environmentamerica.org/texas/
    articles/texas-and-global-warming-where-are-greenhouse-gases-
    coming-from/

12. https://poweralliance.org/2024/01/22/opinion-messer-the-texas-
    grid-did-its-job-during-the-freeze-but-the-jobs-not-done/

13. https://climatecommunication.yale.edu/publications/climate-
    change-in-the-american-mind-politics-policy-spring-2024/

CHAPTER 9: WE CAN AND MUST ACT: IT'S UP TO US

1. https://www.iea.org/reports/methane-tracker-2021/methane-
    and-climate-change

2. Office of Sustainability, https://www.austintexas.gov/department/
    sustainability

3. https://www.epa.gov/sustainable-management-food/wasted-
    food-scale

4. https://www.whitehouse.gov/briefing-room/statements-releases/
    2021/ 04/22/fact-sheet-president-biden-sets-2030-greenhouse-
    gas-pollution-reduction-target-aimed-at-creating-good-paying-
    union-jobs-and-securing-u-s-leadership-on-clean-energy-
    technologies/Carbon Capture - Rebate instead tax credit.

5. Rhodium Group, https://rhg.com/research/climate-clean-energy-
    inflation-reduction-act/

6. CarbonBrief, Daily Briefing, January 29, 2024, https://www.
    carbonbrief.org/daily-brief/china-added-more-solar-panels-in-
    2023-than-us-did-in-its-entire-history/

7. Todd, Thomas, (March 27, 2023) *HATCH* Article, https://www.hatch.com/About-Us/Publications/Blogs/2023/03/Repurposing-coal-fired-power-plants-benefits-and-challenges

8. Yale Program on Climate Change Communications: https://climate communication.yale.edu/publications/global-warmings-four-indias-2022-an-audience-segmentation-analysis/

9. High Schoolers for Carbon Dividends, https://www.hs4cd.org/

10. Students for Carbon Dividends, https://www.s4cd.org/

APPENDICES

APPENDIX 1

RESOLUTION NO. <u>20071129-045</u>

WHEREAS, the City of Austin is recognized as an international leader in sustainable building practices; and

WHEREAS, sustainable building practices conserve energy, water and other natural resources, promote human health and safety, create high-quality and enduring structures, enhance economic value and reduce costs over the life of a building; and

WHEREAS, construction, operation and maintenance of buildings in accordance with the United States Green Building Council's (USGBC) Leadership in Energy and Environmental Design (LEED™) guidelines promotes these goals; and

WHEREAS, the City of Austin has a resolution number 000608-43 addressing sustainable building practices and requiring LEED™ Silver certification for new construction of municipal buildings; and

WHEREAS, the proposed resolution seeks to clarify and expand upon the sustainable building practices established in resolution number 000608-43, particularly with regard to the operation and maintenance of existing municipal buildings; and

WHEREAS, the City of Austin reaffirms its commitment to community sustainability and to reducing the cost of municipal operations to City taxpayers; **NOW, THEREFORE,**

BE IT RESOLVED BY THE CITY COUNCIL OF THE CITY OF AUSTIN:

The City Council directs the City Manager to take the steps necessary to assure that goals, standards and criteria for achieving the highest optimal outcomes for sustainability in municipal projects are implemented in accordance with this Resolution.

BE IT FURTHER RESOLVED:

That the Council directs the City Manager to:

1. For new municipal buildings and for major renovations and additions, develop criteria to assess and achieve the highest optimal levels of sustainability using the appropriate LEED™ rating tool, with a policy of achieving, at a minimum, LEED™ Silver rating certification for all building projects meeting the following scope and budget criteria:

   a. The scope includes work in each of the five major LEED™ checklist categories of: sustainable sites, water efficiency, energy and atmosphere, materials and resources, and indoor environmental quality, or other such categories as may be added or changed by the USGBC from time to time.

   b. The construction cost of the project is at least $2,000,000.

2. Develop Baseline Sustainability Standards for: energy efficiency, water conservation, water quality and storm water management, construction waste management, and locally sourced materials as defined by Austin Energy Green Building.

3. Develop best-practices design criteria for achieving Baseline Sustainability Standards in projects that do not meet the scope and budget criteria set forth above for new municipal buildings and for major renovations and additions·

    a. Using the LEED™ check list, or an alternative rating system approved by the City Manager, as guidelines for design criteria; and
    b. Meeting or exceeding applicable Baseline Sustainability Standards.

4. For municipal building renovations, additions and interior finish-out projects, develop criteria to assess and achieve the highest optimal levels of sustainability using the appropriate LEED™ rating tool, with a policy of achieving, at a minimum, LEED™ Silver rating certification for all projects meeting the following scope and budget criteria:

    a. The scope includes work in each of the three major LEED™ checklist categories of: energy and atmosphere, materials and resources, and indoor environmental quality, or other such categories as may be added or changed by the USGBC from time to time.

b.   The construction cost of the project is at least $300,000.

5.   Develop best-practices design criteria for achieving Baseline Sustainability Standards in projects that do not meet the scope and budget criteria set forth above for renovations, additions and interior finish-out projects:

a.   Using the LEED™ check list, or an alternative rating system approved by the City Manager, as guidelines for design criteria; and

b.   Meeting or exceeding applicable Baseline Sustainability Standards.

6.   Develop protocols for achieving the highest optimal levels of sustainability in existing municipal buildings and facilities, including:

a.   Develop and implement policies for whole building operations including cleaning/maintenance, recycling programs, energy-use monitoring and reduction of building energy use in accordance with City of Austin Administrative Bulletin 05-01 (Designation of Energy Manager and Establishment of Energy Efficiency Policy), monitoring and reduction of water and materials use, and improvement of indoor environmental quality.

b.   Develop and implement criteria and policies for commissioning and re-commissioning to facilitate optimal performance of the facility throughout the building life.

c.   Develop and implement guidelines and policies for sustainability upgrades over time for building operations, systems and minor space use modifications.

d.   Develop and implement policies for exterior building site maintenance, including: reduction and elimination of greenhouse gas and ozone-forming emissions associated with landscape equipment and other activities; efficient/optimized use of water and energy; purchasing environmentally preferred products and waste stream management.

e.   Provide on-going training for building operations and maintenance staff on achieving and maintaining optimal building performance.

f.   Provide on-going training for building users to optimize building performance.

g.   Develop purchasing policies and guidelines to facilitate optimal building performance.

h.   Require annual reports to Council and City Manager detailing: compliance with operation and maintenance polices, guidelines and criteria; costs and savings associated with compliance; and highlighting policy and budget recommendations for advancement of sustainability goals.

7. Develop and secure staffing and budget resources to fully
    implement the above policies, including:

    a. Identify and secure funding to support achievement of
        LEED™ certification and Baseline Sustainability
        Standards. Implement budget and accounting measures to
        ensure that lifecycle savings from sustainability measures
        are incorporated into capital and operation and
        maintenance budget decision-making processes.

    b. Identify and develop staffing resources to perform all
        duties necessary for LEED™ certification and for the
        implementation of Baseline Sustainability Standards in
        municipal projects. Ensure that staff resources and
        expertise are sufficient such that consultant services for
        LEED™ certification and other sustainability priorities are
        necessary only in exceptional circumstances.

    c. Develop a consultant rotation list for energy modeling
        while developing staff-level resources and expertise in
        energy modeling such that, beginning in Fiscal Year 09,
        consultant services are necessary only in exceptional
        circumstances.

    d. Identify interdepartmental resources currently working on
        municipal sustainable building issues and establish clear
        roles and responsibilities.

8.   Establish oversight and communication mechanisms to ensure efficient and comprehensive implementation of the policies outlined above, including:

   a.   Establish a standing interdepartmental sustainability working group to develop, review, implement and oversee municipal project best practices.

   b.   Establish a review process for municipal projects to ensure a consistent approach to planning and budgeting, including analysis of lifecycle costs.

   c.   Develop a public communications plan, including a website to provide information about City building projects, highlight sustainable building achievements and share information regarding best practices and performance results.

   d.   Report annually to Council and the City Manager on performance measures and outcomes for achieving the above policies and goals.

9.   This resolution replaces and supersedes Resolution No. 000608-43.

ADOPTED: ___November 29__, 2007   **ATTEST:** ____________________
Shirley A. Gentry
City Clerk

APPENDIX 2

RESOLUTION NO. <u>20210902-042</u>

WHEREAS, in June of 2019, Council approved Resolution No. 20190619-091, which required robust labor protections and sustainability requirements for third party development agreements that occur on city-owned land; and

WHEREAS, in addition, the Resolution directed the City Manager to determine how the City could implement the goal of requiring all developments on city-owned land to create zero waste, net zero energy, and net positive water buildings and to recommend updates to the existing Green Building policy (Resolution Number 20071129-045) to further consider appropriate thresholds and policy revisions for public-private partnership projects (P3s's), minor renovations, and leased spaces; and

WHEREAS, since the resolution's passage, a cross-departmental team including representatives from the Public Works Department, Office of Sustainability, Real Estate Services, Capital Contracting Office, Purchasing Office, Economic Development Department, Austin Energy Green Building, Small and Minority Business Resources, Aviation, Building Services, Parks and Recreation Department, Austin Convention Center, Austin Public Library, and the internal Strategic Facilities Governance Team have worked to craft recommendations; and

WHEREAS, staff responded to Resolution No. 20190619-091 with a June 12, 2020, staff memo recommending multiple policy changes to bolster the City's existing green building policies; and

WHEREAS, the proposed changes outline a set of key policy priorities related to site selection guidance, net-zero and low carbon guidance, health and wellness guidance through the WELL building standards, and sustainable landscape guidance; and

WHEREAS, the memo further recommends updated performance standards for new construction for capital improvement projects, major and minor renovations and interior finish-outs for capital improvement projects, third-party financed and alternative delivery projects (such as public-private partnerships), and proposed new requirements for leased spaces; and

WHEREAS, the memo also recommends proposed requirements for new construction capital improvement projects, major and minor renovations and interior finish-outs for capital improvement projects, and third-party financed and alternative delivery projects (such as public-private partnerships); these proposed requirements include mandatory feasibility analyses for rooftop solar installation, avoidance of natural gas, use of auxiliary water, provision of EV charging stations, and mandatory water balance calculations; and

WHEREAS, staff recommended that further policy discussion take place regarding WELL and other building standards; updates to existing facilities; and requirements for public art; and

WHEREAS, the June 12, 2020, staff memo details these recommended changes and considerations; and

WHEREAS, staff has continued refining proposed changes to the Green Building Policy and provided Council with updated recommendations in the summary matrix and detailed policy language dated May 25, 2021 ("Exhibit A"); and

WHEREAS, the City of Austin affirms its commitment to reducing its carbon footprint, reducing water use, and fostering a community well-poised to address climate change and mitigate environmental pollution; **NOW, THEREFORE,**

BE IT RESOLVED BY THE CITY COUNCIL OF THE CITY OF AUSTIN:

1. The Council approves the proposed updates to the City of Austin Green Building Policy contained in Exhibit A of this resolution.
2. The Council directs the City Manager to continue conversations with relevant boards and commissions regarding the outstanding issues outlined in the June 12, 2020, memo and to return to Council at a work session prior to December 9, 2021, so that Council may provide additional feedback or direction.
3. This resolution amends Resolution No. 20071129-045.

ADOPTED: _____September 2_____, 2021 **ATTEST:** _______________________
Jannette S. Goodall
City Clerk

APPENDIX 3

CONCRETE RESOLUTION - 20230420-024

RESOLUTION NO. 20230420-024

WHEREAS, the City of Austin has long been at the forefront of combating climate crisis by creating policies that reduce carbon emissions, and improve the environment and quality of life for residents; and

WHEREAS, the City, as a leader in innovation, routinely identifies and tests solutions to complex challenges facing the City; and

WHEREAS, the Austin Climate Equity Plan includes goals of equitably reaching net-zero community-wide greenhouse gas emissions by 2040; and

WHEREAS, in 2019 the Environmental Commission passed Motion 20190619-007c related to piloting low-carbon concrete; and

WHEREAS, Austin Energy's Green Building Program has encouraged the use of Environmental Product Declarations to increase understanding of the environmental impact of the products and materials used by the City; and

WHEREAS, the federal government has defined concrete as one of the fou highest contributing construction materials to greenhouse gas (GHG) emissions along with asphalt, steel, and sheet glass; and

WHEREAS, concrete plays a vital role in our daily lives in shaping the bui environment around us, from schools, hospitals, and housing, to roads, bridges, tunnels, runways, and sewage systems; and

WHEREAS, if concrete were a country, it would be the third largest emitte of greenhouse gases on earth, behind China and the United States; and

WHEREAS, concrete is a complex recipe of materials with a diverse set of uses and applications; and

WHEREAS, cement, the key ingredient that gives concrete its strength is produced by baking limestone in kilns, a process that typically uses coal or natural gas as fuel and consumes a large amount of energy while releasing carbon dioxide (CO_2) from the combustion; and

WHEREAS, cement production accounts for seven percent of all global carbon emissions, more than three times the emissions produced by aviation; and

WHEREAS, technology exists to enhance the sustainability of concrete mixtures that will enable concrete producers to effectively reduce GHG emissions resulting in economic and climate benefits; and

WHEREAS, the transition to more sustainable concrete has already begun in Austin with more than 1,000,000 cubic yards of lower carbon concrete used since 2019 with no net additional cost to the marketplace and reducing the carbon footprint of these materials by thousands of tons of CO_2; **NOW, THEREFORE,**

BE IT RESOLVED BY THE CITY COUNCIL OF THE CITY OF AUSTIN:

The City Manager is directed to explore a plan to transition all future City contracts and projects to low-embodied-carbon concrete. The plan should address the following:

1. Create procedures for tracking concrete that:
   - Identify how much concrete is used on City projects, and projects that will be owned and maintained by the City to better understand Austin's environmental footprint and influence our future designs, specifications, and decisions;

- o Require submittal of Environmental Product Declarations by concrete producers to encourage and influence more sustainable concrete production; and
- o Develop a strategy, process, and schedule for City staff to review, comment, pilot, and approve alternative mix designs proposed by local concrete producers.

2. Establish a standard and/or definition for low-embodied-carbon concrete that may include one or more of the following strategies, but not limited to CO_2-injected concrete allowing for reduced cement use, reducing cement used in concrete by limiting the concrete specified to having only the characteristics needed for a reliable design life and performance, extensive use of more blended cement and Supplementary Cementitious Materials, and potential for using Performance Engineered Mixtures.

3. Provide City Council with an annual report on the progress of alternative mix designs.

BE IT FURTHER RESOLVED:

The City Manager is directed to return to Council with a plan and implementation schedule no later than November 30, 2023.

ADOPTED: _____ April 20 _____, 2023 **ATTEST:** _______________

Myrna Rios
City Clerk

APPENDIX 4

NAMES OF ENVIRONMENTAL GROUPS

Citizens' Climate Lobby - https://citizensclimatelobby.org/

Sierra Club - https://www.sierraclub.org/

Greenpeace - https://www.greenpeace.org/usa/

Earthjustice - https://earthjustice.org/

Environmental Voter Project - https://www.environmentalvoter.org/

Environmental Defense Fund - edf.org

Friends of the Earth - foei.org

Project Drawdown - http://www.drawdown.org

The National Audubon Society - audubon.org

The Natural Resources Defense Council - nrdc.org

The Nature Conservancy - https://www.nature.org/en-us/

Climate Justice - http://www.350.org

Center for Health Environment and Justice - https://chej.org/

EN-ROADS Climate Interactive - https://www.climateinteractive.org/en-roads/

Environment America - https://environmentamerica.org/

The Climate Reality Project - https://www.climaterealityproject.org/

Water for People - https://www.waterforpeople.org

Rainforest Alliance - http://www.rainforest-alliance.org

American Forests - http://www.americanforests.org

Conservation International - http://www.conservation.org

WeForest - http://www.weforest.org

Jane Goodall Institute - http://www.janegoodall.org

Ocean Conservation - http://www.5gyres.org

Oceana - Protecting the World Oceans - http://oceana.org

Sustainable Harvest International - http://www.sustainableharves.org

Earth Guardians - http://www.earthguardians.org

Cool Effect - http://www.cooleffect.org

"MOTHER EARTH TRILOGY" BY SWADESH M. MAHAJAN, PROFESSOR OF PHYSICS AT UT AUSTIN

Poem 1: Indulgent Mother and the Wayward Child, April 28, 2019

Do not ignore my screams of agony

I am suffocating, my wayward child

The best fruit of my womb so lush

unremitting toil, striving, willing it took

Long, indeed, was the gestation

Profound, indeed, were changes to make

a transformation so awesome, so full

From a dead rock to a life-brimming lake

I did not exactly bear you in my womb

But I dreamed of you and crafted you

Gave you abundance, the gift to grow

Gave you sentience, the power to feel

Gave you curiosity and power to create

Held your hand and blessed your creed

You were, indeed, a perfect child
This is, really, what I thought
You loved me and called me Mother
Worshipped me as a giver of plenty
husbanded wisely my vast lush domains
took some and gave some
who could imagine Earth without man!

You were special but not the only one
the plants, the animals, big and small
were just as precious, my ornaments
the father sun's radiant blessings
gave me beauty, my bounty, my wealth
my love and generosity sustained you all

Then one day, the wise amongst you
declared man's dominion over me
not just the plants, the animals few
but over the rocks and the oceans blue
your learned sages spread the lore
that a divine power has willed it so

It took a while, but the idea spread
Earth is to be exploited to enrich the man
his needs, wastefulness, luxuries unbound
greed and arrogance the dominant theme

unlimited perpetual growth rose in ranks
of divinities that fully control the clan

You dug deep into my body, my womb
burnt the wealth that took aeons to make
polluted my surroundings my breathing space
that I am suffocating you refuse to heed
some did express concern as my fever rose
the rest dug deeper driven by lust and greed

The creativity the drive that made you special
the shining jewel that I could boast
An object did I become in your hands
To use to control to exploit with limitless haste
the goose that once laid the golden eggs
is heading to turn into fevered waste

I am loving and I am generous
I could forgive the transgressions
the violent incursions into my body and soul
I could and would indulge the crimes
that in selfish greed you did commit
but my fever I cannot control

Only you and only you
by an overt act of will and courage
can slay the dragon that robs me

of health, of beauty, of balance,
of equanimity, of power to make you bloom
reaching the heights transcending a dream

I will always be there despite fever
Perhaps cold and frigid in parts
storms and hurricanes may pummel my body
my swelling breasts may get dry

But it is your fate, my precious child
that fills me with fear, makes me cry
Wake up then, I beg of you
Do not ignore my screams of agony
I am suffocating, my wayward child

Poem 2: A Flash Back, January 24, 2021

It is not that in the days of yore
I had a perfect ambience
Nothing natural could
My complexion wasn't golden green
majestic though my body
it was laced with lumps and blisters
turbulent winds and raging storms
not serenity were the likely norms

Who knows what happened then

There was no word, no images

No thoughts no memory

There was no Eden 'til you came

There was no conception of fear or joy

Neither of right and wrong

Before your wonted awareness

It was all nescience innocence complete

I had no name then no identity

no narrative of what I am

how was I born from which loins

which womb did bear me so long

what moved me kept me together

why was I there to begin with

mere succession of events endless

natural purposeless relentless

Behold the event so singular in extremis

Not the violent frightening thunder

not the mighty river breaking its banks

not jutting out of the mountain yonder

not even the replicating first life

episodic was your first arrival

It was your sentience sui generis, so unique

that gave me a name, I became Earth.

I was your invention transcending discovery
experiencing random scattered parts of me
You arranged them in a tapestry
You chiseled and carved and painted
In fear, joy, wonder and amazement
With humility with confidence supreme
You sculpted me into a form sublime
Goddess mother the dominant theme

Yes - I am terra-earth the soil to plough
I am Tontazin the personal mother
Pachamama to love worship and bow
Gaia too, both the old and new
Vasundara whence arose all worlds
And to which all creation will finally go
my wholesomeness my infinite reach
your mind did fathom your heart did feel

You gave me a name, you are my father
You are my mother–loving cherishing
You created living earth from rock and water
Harnessed me into a horn of plenty
I fed you and nourished you in turn
Father changed into my beautiful child
the transition was easy, seamless
so natural destined to last for ever

A worshipper, a child become an exploiter

your boosted strength power limitless greed

mocks at what I gave of my easy will

you broke the sacred covenant

my perpetual agony and thy mounting guilt

must follow thy heartbreaking heartless deed

Grudge me not, then, the words I ply

The thoughts I think the feelings I feel

I know not how with past may I deal

My discontent with the present full of agony

My fears and hopes for the days to come

My eloquence that may run afoul

You, you alone are the author of it all

But are you ready to face thy inner soul!

Poem 3 : Back to the Present, January 28, 2021

Here I go again groaning in pain, pain

that grows by the day intense by the night the coldhearted
neither hear nor care

But where are those who know my plight

Did you watch me cough choking on dust Sweating with heat,
weakened by fever Losing control over my air my water
Lashed by storms that never end Untimely rain with floods
so ruinous

dry spells and forest fires

burning through the green luxuriance of thy mother earth,
    nurse so generous

you see it all, that is the truth

but do you notice, do you feel

does it disturb your daily dallyings

A sleepless night may be scary dreams a tear or two for my
    aches and pains did you fast did you even seek benedictions
    from the Gods you pray that my raging fever may melt away

The learned amongst you who care Make models of my state-
    my agony With much patience and matching travail brilliant
    minds calculate my destiny

Dire are the predictions

Hotter and sicker will be my fate

If not checked with vigor and haste

In little time it may be far too late

Some believe some ridicule

Some rejoice in the coming regime behold the tundra bloom

Economic boom they scream

Shipping lanes in the arctic

Fruit orchards in Siberia

Shopping malls north of stream alluring beyond their wildest
    dream

some close their eyes refuse to see some see yet take no action

there is not much I can do they decree other wise men, full of
    knowledge bereft of sense but loudly proclaim

it is too late to reverse the speeding train

In this veritable Greek tragedy

Doomed on earth is man's domain

The story of the powerful, the aware They often gather, collectively declare With much hoopla much fanfare

Rio declaration topped by Paris Accord I do believe they really care

They plan too little with achievements rare Hark the anguished cries of the teen Responsible brave with courage to dare

Emergency care is what I sorely need Watch my burning body my heart bleed Phase out all business that makes me ill Speed is of essence with compromise nill business as usual will spell my doom Only a living me can make you bloom You know it all my child so bright

Only fast action can make it right

Arise, awake, my narrator, my son Do not ignore my screams of agony You and I have a shared destiny

C

D

E

ABOUT ATMOSPHERE PRESS

Founded in 2015, Atmosphere Press was built on the principles of Honesty, Transparency, Professionalism, Kindness, and Making Your Book Awesome. As an ethical and author-friendly hybrid press, we stay true to that founding mission today.

If you're a reader, enter our giveaway for a free book here:

SCAN TO ENTER
BOOK GIVEAWAY

If you're a writer, submit your manuscript for consideration here:

SCAN TO SUBMIT
MANUSCRIPT

And always feel free to visit Atmosphere Press and our authors online at atmospherepress.com. See you there soon!

ABOUT THE AUTHOR

KALPANA SUTARIA is an architect and a project manager working for the City of Austin. She has worked on climate responsive buildings and has promoted sustainable guidelines in the city's buildings throughout her career. She has been a volunteer at the Citizens' Climate Lobby, which is working to create a political will to transition away from polluting fuels. She believes that green buildings are crucial, but to cool our environment, we need policies to lower carbon from all sectors of our economy. She has a master's in architecture from the University of Texas at Austin. She is a licensed architect (AIA) in the State of Texas, LEED Accredited Professional (CD+C), and a Project Management Professional (PMP).